T0297769

BECOMING MIT

BECOMING MIT

MOMENTS OF DECISION

EDITED BY DAVID KAISER

THE MIT PRESS
CAMBRIDGE, MASSACHUSETTS
LONDON, ENGLAND

First MIT Press paperback edition, 2012

© 2010 Massachusetts Institute of Technology

This book was set in Engravers Gothic and Bembo by Toppan Best-set Premedia Limited.

Library of Congress Cataloging-in-Publication Data

Becoming MIT : moments of decision / edited by David Kaiser.
   p.   cm.
Includes bibliographical references and index.
ISBN 978-0-262-11323-6 (hc. : alk. paper)—978-0-262-51815-4 (pb. : alk. paper)
1. Massachusetts Institute of Technology—History. I. Kaiser, David.
T171.M428B43   2010
378.744'4—dc22

2009052641

CONTENTS

ACKNOWLEDGMENTS

It is a pleasure to thank Susan Hockfield for inviting me to join the original planning committee for the 150th anniversary of the Massachusetts Institute of Technology (MIT). Kathryn Willmore, who chaired that committee, championed the idea of this book and helped get it off the ground. I am grateful to the chapter authors for their unstinting enthusiasm and continued good spirit, even in the face of multiple revisions. Joanne Barkan provided superb editing assistance, above and beyond the call of duty. I am indebted as well to Chihyung Jeon, Deborah Douglas, and Frank Conahan for their expert assistance with the images. Over the course of this project, Kathryn Willmore and David Mindell provided financial support from MIT150 funds. Finally, I want to extend my thanks to Marguerite Avery and Ellen Faran at the MIT Press for their steadfast support of the project.

BECOMING MIT

INTRODUCTION: MOMENTS OF DECISION

DAVID KAISER

The Massachusetts Institute of Technology celebrates its 150th anniversary in 2011. The Institute that we know today—a world leader in science and technology, innovation and education—evolved along a twisting path, its course buffeted by decisions made within MIT and by changing conditions beyond it. The sesquicentennial provides an occasion to take stock of MIT's history, to pause, look back, and reflect. It is a time to examine how MIT has exemplified and transformed broader trends within higher education, and look forward to the challenges and opportunities that the Institute will face in the coming years.

MIT has always been a forward-looking place, rarely dwelling on its past. Nonetheless, it is easy—and on such occasions, appropriate—to celebrate MIT's enormous achievements over its still-young history. Time and time again, MIT faculty and students have championed the Institute's motto, "Mens et Manus," mind and hand. Their research has expanded our understanding of everything from the smallest bits of matter to the grandest structures in the cosmos, from the basic building blocks of life to the most complex features of economics and society. MIT can boast 154 members of the National Academy of Engineering, 160 members of the National Academy of Sciences, 50 Nobel laureates, 33 MacArthur "genius award" fellows, and even four Pulitzer Prize winners among its faculty, staff, and alumni—and counting.[1] Throughout its history, MIT has rallied to the nation's defense, most famously during World War II and the Cold War. From the start MIT has produced leaders in pedagogical innovation, entrepreneurship, science policy, and more. For 150 years it has been an engine, translating esoteric research into the tools with which we lead our lives every day.

Consider, for example, the areas of communications and computing. Alexander Graham Bell pursued much of his research on what became the telephone in MIT's fledgling physics laboratory in the 1870s. During the middle decades of the twentieth century, MIT researchers invented such fields as information theory, cybernetics, and artificial intelligence. They transformed the latest ideas about silicon chips, digital computation, time-shared computing, and distributed networks from scratch pads to real-world devices. The Internet and many of its most important features, such as encryption technologies that allow everything from chatty emails to high-level bank transfers to flow smoothly around the world, all have thick MIT ties. Today MIT is home to pioneering research in quantum computing, the next new wave in processing and sharing information.[2]

The study of aeronautics at MIT likewise has a long and storied past. An MIT student designed and built a functioning wind tunnel as early as 1896, five years before the Wright brothers enlisted a similar technology to design their famous first aircraft. Aero/astro maven Charles Stark Draper and his crew later designed the guidance and navigation systems for everything from ballistic missiles to the Apollo moon landings. Several alumni of the Department of Aeronautics and Astronautics have served in top positions in the National Aeronautics and Space Administration (NASA). To date, alumni from the department have participated in more than one-third of NASA's manned space flights, accruing more than ten thousand hours in space.[3]

Biologists at MIT were quick to answer President Richard Nixon's call in the early 1970s for a "war on cancer." Their efforts paved the way for the burgeoning biotechnology industry, with many of the most important firms clustering around MIT. More recently, MIT investigators spear-headed the massive Human Genome Project of the National Institutes of Health (NIH), contributing to the first complete sequencing of the human genome in 2003. Alongside these publicly financed activities, thirty years of private investment in the life sciences at MIT has brought an explosion of cutting-edge facilities: from the Whitehead Institute for Biomedical Research (founded in 1982), to the McGovern Institute for Brain Research (2001), the Picower Institute for Learning and Memory (2002), the Broad Institute of MIT and Harvard for research in biomedicine (2004), and the David H. Koch Institute for Integrative Cancer Research (2010).[4]

From Noam Chomsky's revolution in linguistics to the pathbreaking insights of Nobel laureates Paul Samuelson, Franco Modigliani, and Robert Solow into markets, international trade, and economic growth, MIT has long been a world leader in the principled study of human cognition and behavior. Intense research across the Institute has worked to transform ideas like these into practical tools, from MIT's famous Media Lab, founded in 1985, to more recent efforts like the Abdul Latif Jameel Poverty Action Laboratory in the Department of Economics, the MIT AgeLab within the Engineering Systems Division, and more.[5]

MIT has led the way in science policy as well. During the 1930s, early in his MIT presidency, Karl Compton chaired the federal Science Advisory Board for President Franklin Roosevelt. Compton's acolyte at MIT, Vannevar Bush, designed and built the new frameworks for science-government partnership during World War II, and helped lay out the postwar path with his famous proposal, *Science: The Endless Frontier*. MIT President James Killian became the first chair of the President's Science Advisory Committee on its founding in the late 1950s. Since that time, MIT has sent a steady stream of advisers to Washington, DC. Professor Sheila Widnall of the Department of Aeronautics and Astronautics, for example, made history as the first female secretary of the U.S. Air Force during the 1990s. And MIT's most recent President Emeritus, Charles M. Vest, now presides over the National Academy of Engineering.[6]

The vision of MIT's founder, William Barton Rogers, is instantiated in various ways in these activities, ranging freely across the abstract and the earthly. Rogers saw the need for a new kind of Institute that could meld sound training in the foundations of natural sciences with hands-on learning in practical arts. He introduced a novel laboratory-based system of instruction with that goal in mind. Thanks to early textbooks like *Physical Manipulation* by MIT physics professor Edward Pickering—published in 1873, the earliest-known physics laboratory manual published in the United States—Rogers's innovative approach became a model for science teaching elsewhere.[7] It has continued to inspire pedagogical innovations at MIT as well, from the Physical Sciences Study Commission (PSSC) in the 1950s, the brainchild of MIT physicist Jerrold Zacharias, to the much-copied Undergraduate Research Opportunity Program (UROP), inaugurated at MIT in 1969.[8] More recent developments include MIT's OpenCourseWare, a bold effort launched in 2001 to make materials from most of MIT's courses freely available to the public via the Internet; and MIT's first-of-its-kind under-graduate major in biological engineering, established in 2005.[9]

This type of MIT training has done more than inspire heady thoughts; it has been good for the bottom line as well. At latest count, the currently active companies founded by MIT alumni—nearly twenty-six thousand companies worldwide—employ roughly 3.3 million people and enjoy annual revenues of two trillion dollars. Together, these MIT-spawned companies operate on a scale comparable to the eleventh-largest economy in the world.[10]

By any measure, then, much has been achieved during these brief 150 years, for which MIT's faculty, administrators, students, and alumni should rightly be proud.

Beyond such litanies of success, commemorations offer an opportunity. Fifty years ago, when the Institute celebrated its centennial, MIT President Julius Stratton hoped that "some day" we might have "a coherent account of the flow of ideas that have influenced academic aims and the methods of teaching at the Institute throughout the century." Such developments, he made clear, "can be understood only when placed in the context of the great industrial and intellectual movements of the times." Stratton looked forward to a careful accounting of the ideas behind the place: of what the Institute has stood for, and what its leaders have hoped it might be.[11]

This book takes up Stratton's challenge. Rather than attempting to recount MIT's history in encyclopedic detail, the chapters that follow interrogate those moments of decision that have helped to define the Institute we know today. Some of these turning points have been dramatic, their import obvious to participants at the time. Others have been more subtle, their full impact only recognizable with hindsight. Moments of intense strife and uncertainty take their place in these pages alongside some of MIT's great successes.

As Stratton had surmised, much can be learned by studying a single institution over a long period of time.[12] MIT's history illuminates some of the sharp shifts in assumptions about the means and ends of higher education, from the Industrial Revolution to our post-industrial society. During that period, the Institute's leaders have held different models in mind—both to emulate and avoid—for shaping a great center of research and teaching. European institutions, especially the budding German research universities, offered one model during the late nineteenth and early twentieth centuries.

Others looked with concern in those years at the land grant colleges and engineering universities popping up across the American West. Often, of course, they looked closer to home. Since MIT's founding, Harvard University has functioned alternately as a rival, doppelgänger, and for a brief moment, degree-conferring partner.

In turn, MIT has long served as a model for other institutions, frequently revealing in extreme form broader trends throughout American university life. For example, many colleges and universities in the United States sought partnerships and patronage from local industries in the 1910s and 1920s; MIT reorganized much of its administration around the idea. Dozens of universities participated in the war effort during the 1940s; MIT secured contracts for defense-related research and development worth roughly $100 million (about $1.2 billion in 2008 dollars), substantially more than any other university and three times more than Western Electric (AT&T), General Electric, RCA, DuPont, and Westinghouse combined.[13] Undergraduate and graduate-level enrollments ballooned after the war across the nation's colleges and universities; MIT churned out record numbers of graduates, topping the list year after year in several fields of science and engineering. By the 1980s, as biotechnology began to flourish and national priorities returned to a renewed emphasis on corporate-university partnerships, MIT again led the way. Meanwhile, in more recent years, MIT and its peer institutions faced vexed questions about equality and discrimination. While other institutions downplayed the problem or invoked fears of litigation to justify inaction, MIT became a role model in the aftermath of its famous report in 1999 on the status of women faculty in the School of Science. Indeed, MIT has served as a model for leading institutions beyond the United States. For decades enterprising institution builders have sought to export the "MIT model" around the world—not always successfully—from Iran to India and beyond.[14] Studying the evolution of MIT thus opens a window on to a much wider educational landscape.

Though invoking the past, anniversary celebrations, like all history, are conducted in the present with an eye to the future. We always strive to craft a usable past—one whose highlights and contours are shaped by our current concerns and anticipations. Such activities must by necessity be selective. There will always be elisions, intended or otherwise. Episodes and individuals that once loomed large in the collective memory inevitably fade from view, our sense of their meaning shading, at times imperceptibly, upon each reconsideration.[15]

When selecting moments from MIT's past for close scrutiny, several of the Institute's present concerns played a role in my thinking. In the shadow of the Cold War, for example, as new security challenges loom, what relationship should MIT forge with the federal government, and what responsibilities should it adopt in and out of wartime? What is the optimal balance between government funding and private investment for research on campus? Meanwhile, after several grinding years of intense curriculum review, the question of what to teach and how best to teach it remains no more settled today than during earlier moments in MIT's past.[16] Much as in William Barton Rogers's day, finding the appropriate admixture of basic science, engineering know-how, and humanistic scholarship that will best prepare MIT's students for careers beyond the Institute presents a continuing challenge. With these large themes in mind, this book brings together scholars with a deep knowledge of the Institute to unpack MIT's tangled and fascinating history.

In chapter 1, Merritt Roe Smith traces Rogers's path toward establishing MIT. Inspired by the maturing of American industries such as railroads, and further informed by his early experiences in geological surveying, Rogers came to value both hands-on learning and political acumen. Even during the difficult years of the American Civil War, Rogers managed to garner state support for his new school. He succeeded by equal parts dogged lobbying and good timing. MIT was among the earliest beneficiaries of the Morrill Land Grant Act, signed into law by President Abraham Lincoln in 1862. The Morrill Act granted state and federal support to higher education in practical subjects. In turn, Rogers parlayed the sizable government funds into an effective fund-raising campaign among private donors, securing the capital he needed to launch MIT.

Protecting a niche for MIT-styled education proved to be no simple task. As Bruce Sinclair describes in chapter 2, prominent voices strove time and again to clarify that MIT's brand of technical training should not become a mere appendage to a "classical" or liberal arts curriculum. Throughout the late nineteenth and early twentieth centuries, MIT's leaders honed and articulated their sense of the Institute with constant reference to nearby Harvard. The comparisons ranged more widely than curricula. Class status was at stake. MIT's champions fought back against perceived snobbishness; their graduates were no mere factory workers or tradesmen. At times they revealed some snobbishness of their own, as when MIT President Francis Walker sniffed in 1893 at well-heeled undergraduates "loafing in academic groves" at (unnamed) liberal arts colleges—a swipe at Harvard that no local readers could miss. Though the Institute struggled through lean financial times and was nearly swallowed up by Harvard on three separate occasions, its move to Cambridge in 1916 marked a significant coming of age.

During the opening decades of the twentieth century, a battle raged for the soul of MIT. In chapter 3, Christophe Lécuyer clarifies the conflicting visions. On one side were faculty members who had completed their doctoral training in Germany during the 1880s and 1890s, and who had returned to the United States imbued with a "pure science" sensibility. A competing faction advocated closer ties to industry. In their eyes, strengthened connections between MIT and industrial corporations would enable MIT to train the next generation of corporate managers. The juggernaut was broken with the selection of Richard Maclaurin as MIT President in 1909. Though trained in mathematical physics, Maclaurin championed the idea of MIT as directly useful to industry—by taking on research tasks at corporations' request, opening up the Institute's libraries to industrial sponsors, and sharing alumni records with corporate recruiters—all in exchange for hefty industrial patronage. Maclaurin's ideas culminated in his "Technology Plan" of 1919, the timing forced by severe budget shortfalls during World War I. Friction mounted throughout the 1920s over who should control the results of research: the professors who conducted the research or the companies that paid for it. Critiques of the Technology Plan grew, some observers fearing that the plan had skewed MIT's efforts toward short-term, applied projects of limited fundamental significance. The pendulum swung back during the 1930s under MIT President Karl Compton, who enjoyed the backing of powerful science-based industrialists at General Electric and Bell Telephone. Curricula were refocused on basic science, department faculty members were reshuffled, and MIT's relationships with industrial patrons were renegotiated to clear a wider space for MIT's intellectual autonomy.

Wartime mobilization jumbled MIT's balance of patronage and autonomy in even more dramatic ways, as Deborah Douglas documents in chapter 4. More than a year before the attack on Pearl Harbor in December

1941, MIT had already dived into defense work. The Radiation Laboratory or "Rad Lab," headquarters of the Allied effort to design short-wavelength radar, mushroomed the quickest. From its humble beginnings in fall 1940— when its entire staff consisted of twenty physicists, three security guards, two stockroom clerks, and a secretary—the lab swelled by war's end to a sprawling complex with a staff of four thousand and an operating budget of one million dollars *per month* (about 12 million dollars per month in 2008 dollars). Dozens of other laboratories on campus followed suit, developing everything from improved gun sights and fuels to synthetic vitamins and new methods of processing and packaging food for military use. Equally important, MIT hosted tens of thousands of enlisted men and military officers as they cycled through special training programs on campus en route to the battlefield. MIT's enormous wartime service transformed its earlier interactions with the federal government, cementing its new role as the largest beneficiary of federal research dollars of any educational institution in U.S. history.

The end of World War II brought no real demobilization to campus. Instead, as I discuss in chapter 5, the hardening of the Cold War sustained the Institute's wartime patterns of operation. Once again, MIT's leaders faced the question of what skills and techniques each student should master—a question exacerbated by the record-breaking rates of growth in both undergraduate and graduate student enrollments. A firehose of federal funding, most of it from the Department of Defense and related agencies, helped launch new laboratories on campus and expand old ones. Even as MIT's research and training came to center around Cold War gadgetry— nuclear reactors, radar receivers, electronic transistors and silicon chips, inertial guidance systems, advanced computers, and more—a blue-ribbon faculty panel asserted that students in the nuclear age also required an appreciation of history, literature, and the patterns of decision making in a democracy. Their arguments led to the founding of MIT's School of Humanities and Social Sciences in 1950, though mastering the right curricular balance remained a challenge throughout the boom years.

During the late 1960s, as the Vietnam War dragged on, tensions erupted at MIT, as Stuart W. Leslie describes in chapter 6. Beyond the sit-ins, marches, and protests that flared at colleges and universities across the United States, the antiwar fervent at MIT elicited prolonged introspection. What should MIT's mission be—as one of the nation's largest defense contractors—during the Vietnam War? The core of MIT seemed up for grabs. Students and faculty debated whether classified research belonged on campus, and whether "socially useful" research had been shunted aside in lieu of large defense contracts. A group of concerned students and faculty called for a "day of reflection"—widely reported in the press as a "strike"— for March 4, 1969. Classes and research were suspended, replaced by teach-ins on scientists' social responsibility. The "March 4th movement" quickly spread to other campuses across the country. Meanwhile MIT's "special laboratories"—Charles Stark Draper's Instrumentation Laboratory and Lincoln Laboratory—became flash points of the new unrest. Devoted largely to defense work, both labs had grown by leaps and bounds during the previous two decades, rivaling the rest of MIT in budget and staff. While an elite committee evaluated whether MIT should divest its ties to the laboratories or redirect the labs' missions, riot police—swinging batons and all—broke up angry protests on the Instrumentation Laboratory's door-steps. "The Sixties" had arrived at MIT.

Questions of scientists' social responsibility lingered even after the dra-matic protests of the late 1960s. In chapter 7, John Durant analyzes how conflicts over scientists' social responsibility became especially acute in the life sciences during the 1970s. MIT biologist David Baltimore teamed up with colleagues around the country in 1974 to alert scientists and policy-makers about the possible dangers of recombinant DNA research. They called for a voluntary moratorium on rDNA work until experts could assess the risks and recommend best procedures. The scientists learned quickly— at times to their dismay—that other groups had strong opinions on the topic as well. Invoking images of Frankenstein's monster run amok, Cam-bridge's colorful mayor, Alfred Vellucci, arranged public hearings on the matter and threatened to ban all rDNA work within city limits. Out of the turmoil emerged an unprecedented mechanism for the public oversight of research at private institutions: the Cambridge Experimentation Review Board, composed of concerned citizens without any particular scientific training. Meeting with scientists and other stakeholders twice a week for more than six months, the board hammered out a set of regulations and guidelines for recombinant DNA research—as it happens, not very different from the original proposals by Baltimore and others. With the new arrange-ments in place, backed by newfound channels of communication along with trust building between town and gown, MIT zoomed to the biotechnology frontier. Private investment poured in, leading to MIT's Whitehead Institute for Biomedical Research as well as several spin-off companies in the Cambridge area.

MIT took on a different kind of oversight in the 1990s. Following four years of careful investigation, the Institute released *A Study on the Status of Women Faculty in Science at MIT* in March 1999. As Lotte Bailyn recounts in chapter 8, the report detailed institutionalized gender discrimination among the faculty in MIT's School of Science. Unfair treatment could be seen in virtually every category, and indeed in anything that could be counted: salaries, laboratory space, teaching assignments, committee loads, and more. The report garnered international attention from the popular press and scholarly reviews alike. Equally newsworthy was MIT's swift response to redress the inequalities and to widen the inquiry to MIT's other schools. More than mere accounting was at stake. The report and its aftermath signaled a broader reconsideration of who fits at MIT; whose contributions are valued; and how best to account for those contributions. In an afterword to the chapter, Nancy Hopkins, who jump-started the process that led to the report, reflects on how life at the Institute has changed since that time.

Finally, in the epilogue, MIT President Susan Hockfield assesses the legacy of MIT's past as it embarks on its future. Amid clear shifts in MIT's disciplinary matrix (heralded most clearly by the booming life sciences on campus and beyond), heated debates over its undergraduate curriculum, and its starkest financial challenge since the Great Depression, MIT once again finds itself confronting significant tests. Gathering again at some future commemoration, MIT's faculty, administrators, students, and alumni will judge how well the Institute has navigated this latest moment of decision.

NOTES

1. Running tallies are maintained at <http://web.mit.edu/ir/pop/awards> (accessed June 22, 2009).

2. Anthony P. French, "Physics Education at MIT: From Bell's Phonautograph to Technology Enhanced Active Learning," *Physics @ MIT* 18 (2005): 40–47; Claude Shannon and Warren Weaver, *The Mathematical Theory of Communication* (Urbana: University of Illinois Press, 1949); Norbert Wiener, *Cybernetics: Or, Control and*

Communication in the Animal and the Machine (Cambridge, MA: MIT Press, 1948); Scientific American, *Information* (San Francisco: W. H. Freeman, 1966); John V. Guttag, ed., *The Electron and the Bit: Electrical Engineering and Computer Science at the Massachusetts Institute of Technology, 1902–2002* (Cambridge, MA: MIT Department of Electrical Engineering and Computer Science, 2005); Isaac Chuang, "Quantum Information: Joining the Foundations of Physics and Computer Science," *Physics @ MIT* 17 (2004): 26–29, 44–45; Seth Lloyd, *Programming the Universe: A Quantum Computer Scientist Takes on the Universe* (New York: Knopf, 2006). See also Steve J. Heims, *John von Neumann and Norbert Wiener: From Mathematics to the Technologies of Life and Death* (Cambridge, MA: MIT Press, 1980); David Mindell, *Between Human and Machine: Feedback, Control, and Computing before Cybernetics* (Baltimore: Johns Hopkins University Press, 2002); Atsushi Akera, *Calculating a Natural World: Scientists, Engineers, and Computers during the Rise of U.S. Cold War Research* (Cambridge, MA: MIT Press, 2007).

3. Lauren Clark and Eric Feron, "A Century of Aerospace Engineering at MIT," in *Aerospace Engineering Education during the First Century of Flight*, ed. Barnes McCormick, Conrad Newberry, and Eric Jumper (Reston, VA: American Institute of Aeronautics and Astronautics, 2004), 31–43. See also Stuart W. Leslie, *The Cold War and American Science: The Military-Industrial-Academic Complex at MIT and Stanford* (New York: Columbia University Press, 1993), chap. 3; David Mindell, *Digital Apollo: Human and Machine in Spaceflight* (Cambridge, MA: MIT Press, 2008).

4. Phillip A. Sharp, "Life Sciences at MIT: A History and Perspective," *MIT Faculty Newsletter* 18, no. 3 (January–February 2006). See also Lynne G. Zucker, Michael R. Darby, and Marilyn B. Brewer, "Intellectual Human Capital and the Birth of U.S. Biotechnology Enterprises," *American Economic Review* 88 (March 1998): 290–306.

5. On the MIT Media Lab, see <http://www.media.mit.edu>; on the MIT Poverty Action Lab, see <http://www.povertyactionlab.com>; on the MIT AgeLab, see <http://web.mit.edu/agelab> (all accessed June 29, 2009).

6. On Compton and the Science Advisory Board, see the chapter by Christophe Lécuyer in this volume. On Bush, see Vannevar Bush, *Science: The Endless Frontier* (Washington, DC: Government Printing Office, 1945); Daniel Kevles, *The Physicists: The History of a Scientific Community in Modern America*, 3rd ed. (Cambridge, MA: Harvard University Press, 1995 [originally published in 1978]), chaps. 21–22; Nathan Reingold, "Vannevar Bush's New Deal for Research: Or, the Triumph of the Old Order," *Historical Studies in the Physical and Biological Sciences* 17 (1987): 299–344. On Killian, see James R. Killian, *The Education of a College President: A Memoir* (Cambridge, MA: MIT Press, 1985); Zuoyue Wang, *In Sputnik's Shadow: The President's Science Advisory Committee and Cold War America* (New Brunswick,

NJ: Rutgers University Press, 2008), chap. 5. On Widnall, see <http://web.mit.edu/aeroastro/www/people/widnall/bio.html> (accessed June 29, 2009). And on Vest, see <http://web.mit.edu/president/communications/profile.html> (accessed June 24, 2009).

7. French, "Physics Education at MIT," 40. Compare Graeme Gooday, "Precision Measurement and the Genesis of Physics Teaching Laboratories in Victorian Britain," *British Journal for the History of Science* 23 (1990): 25–51; Kathryn Olesko, "The Foundation of a Canon: Kohlrausch's *Practical Physics*," in *Pedagogy and the Practice of Science: Historical and Contemporary Perspectives*, ed. David Kaiser (Cambridge, MA: MIT Press, 2005), 323–356.

8. On the Physical Sciences Study Commission, see John Rudolph, *Scientists in the Classroom: The Cold War Reconstruction of American Science Education* (New York: Palgrave, 2002). On MIT's Undergraduate Research Opportunity Program, see <http://web.mit.edu/urop/basicinfo> (accessed June 29, 2009).

9. On OpenCourseWare, see "Auditing Classes at M.I.T., on the Web and Free," *New York Times*, April 3, 2001; Steven Lerman, Shigeru Miyagawa, and Anne H. Margulies, "OpenCourseWare: Building a Culture of Sharing," in *Opening Up Education: The Collective Advancement of Education through Open Technology, Open Content, and Open Knowledge*, ed. Toru Iiyoshi and M. S. Vijay Kumar (Cambridge, MA: MIT Press, 2008), chap. 14. On the new biological engineering major, see Gareth Cook, "Revolutionary Major Set to Be Born: Biological Engineering to Be First Field Created by School in 29 Years," *Boston Globe*, February 16, 2005.

10. Edward B. Roberts and Charles Eesley, *Entrepreneurial Impact: The Role of MIT*, Kauffman Foundation report (February 2009), available at <http://web.mit.edu/newsoffice/kauffman.html> (accessed June 29, 2009)

11. Julius Stratton, "Report of the President," in *MIT Report to the President* (1961), on 5. Most MIT *Reports to the President* are now available online at <http://libraries.mit.edu/archives/mithistory/presidents-reports.html>. Here and throughout the book, these will be cited by year as *MIT Annual Report*.

12. Two of my favorite examples include Rebecca Lowen, *Creating the Cold War University: The Transformation of Stanford* (Berkeley: University of California Press, 1997); Andrew Warwick, *Masters of Theory: Cambridge and the Rise of Mathematical Physics* (Chicago: University of Chicago Press, 2003). See also the witty and insightful study by William Clark, *Academic Charisma and the Origins of the Research University* (Chicago: University of Chicago Press, 2006).

13. Karl Compton, *MIT Annual Report* (1945), on 8. See also Leslie, *The Cold War and American Science*, 6.

14. Lowen, *Creating the Cold War University*; Leslie, *The Cold War and American Science*. See also Roger Geiger, *Research and Relevant Knowledge: American Research Universities since World War II* (New York: Oxford University Press, 1993). On international developments, see Stuart W. Leslie and Robert Kargon, "Exporting MIT: Science, Technology, and Nation-Building in India and Iran," *Osiris* 21 (2006): 110–130.

15. Pnina Abir-Am, ed., *Commemorative Practices in Science: Historical Perspectives on the Politics of Collective Memory*, published as *Osiris* 14 (1999): 1–383. See also Maurice Halbwachs, *On Collective Memory*, trans. Lewis A. Coser (Chicago: University of Chicago Press, 1992).

16. Natasha Plotkin, "In the Wake of GIR Defeat, Back to the Drawing Board," *The Tech* 129, no. 6 (February 20, 2009). See also Rosalind Williams, *Retooling: A Historian Confronts Technological Change* (Cambridge, MA: MIT Press, 2003).

MERRITT ROE SMITH

On July 2, 1864, one of MIT's early benefactors, Dr. William Johnson Walker, penned the words, "God Speed the Institute."[1] Walker, a retired Boston surgeon whose wealth derived from astute investments in local industrial ventures, was strongly attracted to MIT founder William Barton Rogers's vision of the new science-based institution. His gift of $60,000 (nearly $850,000 in 2008 dollars) arrived at a critical moment. MIT was already three years old, but it had yet to assemble a faculty or begin classes.[2] The problem was a lack of financial resources: hard cash was a scarce commodity in the midst of the bloody American Civil War. Yet Rogers and his circle of supporters—local merchants, engineers, and academicians— persisted, intent on getting the new "polytechnic school" up and running as soon as possible. That day arrived on February 20, 1865, less than two months before the war ended, when fifteen students began classes in Boston's Mercantile Building on Summer Street.[3] Rogers's dream had come true, but it was a long time—thirty-six years—in the making.

THE EDUCATION OF AN EDUCATOR

Born in 1804, Rogers was raised in an academic family, studied at William and Mary College in Williamsburg, Virginia, where his father, Patrick, taught, and gravitated toward the new field of geology during the late 1820s and 1830s.[4] Those were years of transformational change in the United States—a time when the young republic was breaking away from its old colonial moorings, and entering a new era defined by factories, steam engines, and railroads.[5] In the words of the British critic Thomas Carlyle, it was a "mechanical age," a historic transitional moment that witnessed the demise of the craft tradition and the rise of factory manufacturing powered by water turbines, steam engines, and automatic machinery.

Far from being uneasy or skeptical about the new technologies, Rogers and his two younger brothers, Henry and Robert, embraced, studied, and gradually constructed their academic careers around them. While teaching in Baltimore in 1828, William Rogers gave a public lecture on railroads, using models borrowed from the newly chartered Baltimore and Ohio (B&O), one of the earliest and most important lines in America. He had become acquainted with Colonel Stephen H. Long, Captain William Gibbs McNeill, and Lieutenant George Washington Whistler, three accomplished army engineers who, under the General Survey Act of 1824, had been assigned by President John Quincy Adams to survey and initiate construction of the privately owned B&O Railroad.[6] Impressed by Rogers's intellectual ability, these officers tried to persuade him that his "ultimate advancement" as a scientific man would be better promoted by entering the field of civil engineering and taking up the "highly respectable and lucrative exertion in the growing spirit for works of internal improvement," notably the construction of railroads, the leading technology of the day. Rogers listened, although in the end he was not willing to abandon an academic career in science for an engineering career in public works. But thanks partly to a "highly gratifying" letter of recommendation from Long and McNeill, Rogers, who was only twenty-three years old, was offered and accepted an appointment to succeed his recently deceased father as professor of natural philosophy and chemistry at William and Mary College.[7]

In summer 1831, Henry and Robert Rogers joined McNeill and Whistler as members of a team engaged in surveying the route of the Boston and Providence Railroad, a project that Robert returned to the following year. The two Rogers brothers were fascinated by the new steam technology and the engineering skills that made railroad construction possible. They informed their elder brother that Captain McNeill had encouraged them to study the geologic features of the landscapes they were surveying. Since William Rogers was shifting his primary research focus to geology, he read their letters with considerable interest, contemplating the connections between pure science and its applications to engineering. Clearly the new technology was having an effect on all three brothers.[8]

For a brief moment, it seemed that Henry and Robert Rogers might actually switch from science to engineering. But they ultimately opted for academic careers in science—Henry in geology, and Robert in chemistry and medicine. As much as the rapidly industrializing nation needed capable engineers, Robert Rogers concluded that opportunities in the new field were limited for persons of his background and training. He lamented to William in January 1833 that "those who have been educated at West Point stand in the way of promotion, and can look forward to certainty of success; they alone are sure of constant occupation in the profession."[9] Robert was correct. Former West Point–trained army engineers dominated the railroad business. As Brown University President Francis Wayland later observed, "The single academy at West Point, graduating annually a smaller number than many of our colleges, has done more towards the construction of railroads than all our one hundred colleges united."[10]

The Rogers brothers' initiation to industry and "internal improvement" did not end with the railroad surveys.[11] Shortly after moving to a new professorship at the University of Virginia in 1835, William Rogers accepted an appointment as superintendent of the state-sponsored Geological Survey of Virginia.[12] Rogers viewed the undertaking as an exciting opportunity—one that would enhance his reputation in his discipline while opening his career to "a much wider field of exertion," as he put it in a letter to his brother Henry.[13]

Yet once the full-scale survey got underway, Rogers began to experience political pressures to produce reports that favored one or another of the competing economic interests within the state. Eastern planters, who controlled the legislature, wanted him to focus on a search for soil-restoring minerals like marl and gypsum in the interest of reviving Virginia's seriously depleted soils. Western Virginians, on the other hand, saw the survey as a chance to identify lucrative coal and iron deposits that would spur the state's lagging industrial development. Rogers recognized these opposing forces and the bitter intrastate political rivalries that accompanied them, but he considered the survey an opportunity to collaborate with his brother Henry, who had accepted a similar position as state geologist of the Commonwealth of Pennsylvania. The brothers intended to correspond and share data as they conducted their surveys with the ultimate objective of formulating a new theory of mountain formation. Yet owing to political pressures in both states that stressed the economic importance of the surveys, the science side of the projects tended to recede into the background.[14]

As it turned out, the Virginia survey proved extremely disappointing. While William Rogers and his team of assistants learned a lot from a purely geologic standpoint (the venture helped to stimulate science in Virginia and propel Rogers's career as a geologist), they were unable to satisfy the carping politicians and economic interest groups that divided eastern from western Virginia. Caught in the maelstrom of Virginia politics, Rogers became the target of political attacks from both sides. "Sneers have been cast upon my labours," he wrote his more fortunate brother Henry in March 1838. "I am at the mercy of the ignorant or the illiberal."[15] Henry eventually published a highly regarded study, *The Geology of Pennsylvania*, but Virginia's legislature twice defeated bills aimed at funding a similar publication by William. "I am utterly tired of waiting upon the movements of the legislature," he confessed to Henry years later. "The lobby working, of which I see a good deal and hear more, is as repugnant to my taste as to my sense of right, and I avoid even the colour of it."[16] Still, he had learned a hard lesson, one that would serve him well in the years ahead when he became involved in lobbying for an institute of technology in Boston.

PURSUING AN IDEA

When William Rogers wrote his comments about the outcome of the state survey, he had already resigned his professorship at the University of Virginia and moved to Boston. A number of factors influenced his decision. To be sure, his unpleasant experience with the survey coupled with frequent student violence at the university (both of which he attributed to the violent nature of Virginia's slave-holding society) encouraged him to look elsewhere. Moreover, he had visited Boston on previous occasions and found the city's "knowledge-seeking spirit" and bustling economic environment highly energizing. This sentiment, coupled with the fact that his brother Henry had moved to Boston in 1844, made the city all the more attractive. Most important of all, Rogers had fallen in love with Emma Savage, a young Bostonian from a prominent family. They married in 1849.[17]

By the time Rogers moved to Boston in 1853, he had already formulated a plan in collaboration with his brother Henry for the establishment of a technical institute in the city. Interestingly, Henry had first mentioned the idea in 1846 when he wrote to William that he was discussing the subject of a "Polytechnic School of the Useful Arts" with John Amory Lowell, the son of the famous textile merchant-manufacturer after whom the city of Lowell was named and founder of Boston's Lowell Institute, a family philanthropy that provided free public lectures for the city's residents.[18] The idea excited William; he was convinced, perhaps more than Henry, that science should occupy a crucial place in the training of architects, engineers, and other "practical men." Drawing on earlier documents that he and his brother had written on the subject, William drafted "A Plan for a Polytechnic School in Boston," which outlined how such a school would be organized and briefly sketched two primary courses of instruction: one laying "a broad and solid foundation" in general physics, chemistry, and mathematics; and the other, "an entirely practical department" that "would enhance instruction in chemical manipulation and the analysis of chemical products, ores, metals, and other materials used in the [practical] arts, as well as of soils and manures."[19] At a time when scientific knowledge was just beginning to be applied to practical purposes in the United States, William Rogers saw and appreciated the connection. According to MIT biologist/historian Samuel Prescott, Rogers's plan of 1846 formed "the embryo of the Massachusetts Institute of Technology."[20]

As the politically tense 1850s wore on in the United States, both William and Henry Rogers continued to build their reputations as professional geologists, presenting papers and lectures on various scientific subjects, including Charles Darwin's revolutionary theory of evolution. Much still remained to be done to convert the general plan for an institution into a functioning reality. Initially William and Henry worked in concert to recruit local educators, merchants, and "leading practical men" to their project. Then the onus of implementing the idea fell to William when Henry accepted a professorship of geology at the University of Glasgow in 1857.

The time for action came in 1859 when William Rogers joined a group interested in public education that was petitioning the state legislature for a grant of land in the newly developed Back Bay area of Boston. Although the petition failed, Rogers refused to abandon the project, and in the midst of the heated presidential election of 1860 that would soon rip apart the Union, he succeeded in getting a revised report calling for the establishment of an "Institute of Technology" approved by a legislative committee. In November 1860, Rogers and a committee of eighteen associates formally

FIGURE I.I
William Barton Rogers, founder of MIT, shown here in 1869.
Source: Courtesy of the MIT Museum.

applied for "an Act of Incorporation of the Massachusetts Institute of Technology," an entity that initially included a museum and a Society of Arts as well as a School of Industrial Science. After undergoing extensive reviews and public discussions, the act passed the legislature and was approved by Governor John A. Andrew on April 10, 1861. Two days later, Confederate forces under General Pierre G. T. Beauregard opened fire on Fort Sumter in Charleston Harbor, South Carolina. The immediate onset of civil war struck an ominous chord for Rogers and his supporters. Rapidly escalating wartime expenditures made it all the more difficult to find resources for the projected Institute. It was the worst possible time to initiate a new educational venture.[21]

Largely because MIT was chartered as a private corporation and has operated as one ever since, historians have downplayed the role that the Commonwealth of Massachusetts played in its establishment and early development. This view is mistaken. From MIT's beginning in 1861 well into the 1890s, state support was critical to the new venture's development and survival. When William Rogers and his supporters originally applied for a charter, they also sought and received a parcel of land in Boston's Back Bay. As a condition of the charter-granting process, however, the legislature also required MIT to raise $100,000 (nearly $2.5 million in 2008 dollars) as a "guaranty fund" endowment within one year. While the Institute found a number of private contributors, it failed to meet the mandated deadline. To avoid risking the loss of its charter, the incorporators petitioned the legislature for an extension in April 1862 and succeeded.[22] But time was still of the essence because it was unlikely that the legislature would grant another extension.

Uncertainty abounded in 1862 and 1863. The war was going poorly for the Union—something that made Rogers's search for financial support all the more difficult. Hard-pressed for funds, Rogers received welcome news when he learned that President Abraham Lincoln had signed the Morrill Land Grant Act on July 2, 1862. The new law stipulated that each state in the Union would receive 30,000 acres of land for each member of its congressional delegation in Washington. In effect that meant that Massachusetts, with two senators and ten representatives, could realize the income from the sale of 360,000 acres of federal land, some of which had to be used for the establishment and support of at least one agricultural and mechanical college within the state.

FIGURE 1.2
MIT students in the classes of 1869 and 1870. Although the war had ended in 1865, their Civil War uniforms and beards were still in style when this photograph was taken in 1869.
Source: Courtesy of the MIT Museum.

Although Republican governor John Andrew sought to use the land grant money to combine MIT and the projected Massachusetts Agricultural College at Amherst with Harvard, William Rogers lobbied strongly against the measure and ended up getting one-third of the land grant income for MIT. With the Institute's formal acceptance of certain conditions made by the Commonwealth (one of which was to offer military training to all its students), MIT became a land grant institution, one of the first in the nation. Equally important, the de facto imprimatur of the state also helped to assure prospective private donors that MIT had a future and was worth supporting. Several key gifts came to the Institute as a consequence, thereby allowing it to complete the guarantee fund by spring 1863. Altogether, $194,588 in land grant funds (roughly $3.8 million in 2008 dollars) came to the Institute between 1865 and 1900. During the same period, MIT would successfully petition the state for added support amounting to another $362,000 (close to $7 million in 2008 dollars). No private donation came close to these amounts. Government aid thus proved to be a critical factor in MIT's establishment and ultimate survival.[23] Indeed, the Commonwealth of Massachusetts became MIT's most important early backer.

JOINING THEORY AND PRACTICE

MIT was not the first polytechnic institute established in the United States. That distinction belongs to the U.S. Military Academy at West Point and its civilian counterpart, the Rensselaer Polytechnic Institute in Troy, New York, founded in 1802 and 1824, respectively. Other science-oriented colleges preceded MIT, most notably the Lawrence Scientific School at Harvard and the Sheffield School at Yale, both founded in the 1840s.[24] William Rogers was familiar with these schools and their curricula when he began to draft his *Scope and Plan of the School of Industrial Science* in 1863. Indeed, his knowledge of their educational objectives, programs, and strengths and weaknesses influenced his vision for the new institute in Boston.

The program of studies that Rogers developed during the 1860s became known as the "New Education." The concept was popularized by Charles W. Eliot, an early MIT chemistry professor and later President of Harvard, but an important strand of the new approach grew out of Rogers's long-standing interest in the useful arts, especially his experience as the director of the Virginia Geological Survey some thirty years earlier. At its core, the New Education sought to combine scientific theory with engineering practice by first introducing students to theoretical principles and then following this with more specialized "practical" subjects that emphasized application to real-world problems. To achieve these goals of scientific breadth and depth, Rogers emphasized hands-on experience through laboratory instruction and experiment. In his view, the laboratory experience brought theory and practice together in a way that far exceeded what could be accomplished by the usual lecture-demonstration-recitation methods commonly employed elsewhere. While other schools (Rensselaer, for example) had laboratories, none stressed laboratory instruction to the extent that Rogers did. Laboratory experience was, for him, the defining feature of his reform-oriented educational agenda. It offered students an unparalleled opportunity to prepare themselves for careers in business, engineering, and industry.[25]

As an educator, Rogers was an eclectic. He drew on many sources in developing his useful arts approach to undergraduate education. The prior experiences of schools in the United States proved instructive, but the most important influences came from Europe. Rogers not only made fact-finding trips to Europe, he also relied on his brother Henry, the Glasgow professor, to supply him with news of the profession, curriculum materials, and other vital pieces of information. Of the European institutions that Rogers visited, he was most impressed with the French academies, especially the Conservatoire des Arts et Metiers and the Ecole Centrale des Arts et Manufactures, both in Paris. The technical school at Karlsruhe, Germany, also received Rogers's favorable attention. French engineering practices, however, proved to be the most significant influence in the formation of MIT's educational program, just as they had with the establishment of West Point a half century earlier.[26]

Although Rogers had been thinking and writing about the importance of laboratory education for years, MIT's laboratory-oriented culture did not emerge quickly. Laboratories, especially their equipment, were expensive, and the financially strapped Institute was operating on a shoestring budget. In fact, the challenge of building well-equipped laboratories persisted well into the 1890s and beyond.

MIT's first laboratory was established in 1867 by Professor Francis H. Storer, a Harvard-educated analytical and industrial chemist who was an early supporter of Rogers's educational philosophy and one of the Institute's first faculty members. In collaboration with his colleague Charles W. Eliot, Storer published *A Manual of Inorganic Chemistry* in 1867 and *A Compendious Manual of Qualitative Chemical Analysis* two years later to supplement their laboratory teaching program. Both texts became widely adopted sources for teaching college chemistry and went through numerous editions. The Storer-Eliot laboratory model quickly became a standard against which other college chemistry programs were measured.[27]

While Storer and Eliot busied themselves with the chemistry lab, Assistant Professor Edward C. Pickering drew up a plan for a physics laboratory. Hired in 1867 to assist President Rogers in teaching physics at MIT, the talented Pickering raced through the ranks, and in 1868 succeeded Rogers as the senior professor of physics when the latter suffered a stroke, stepped

FIGURE 1.3
The Rogers Laboratory of Physics at MIT, ca. 1869. MIT helped inaugurate new laboratory-based techniques of instruction soon after its founding.
Source: Courtesy of the MIT Museum.

down from the presidency, and took an extended leave from the Institute.[28] The mathematician John D. Runkle, Rogers's closest colleague and successor as President of MIT, thought highly of Pickering. "I am more & more impressed with the great treasure we have in him," he wrote Rogers's wife, Emma, in April 1869. "I am convinced that in time we shall revolutionize the instruction of Physics just as has been done in Chemistry."[29] With the MIT Corporation's approval, Pickering had the new lab up and running by fall 1869.

Pickering's laboratory proved a great success. By spring 1870, he reported with considerable pride that sixty students were working in the lab and getting "a far more practical knowledge of physics than [could be acquired] by devoting the same time to lectures."[30] Not only were students performing experiments under Pickering's watchful eye and gaining useful knowledge as they proceeded, they also were doing original research. In June

1870, for instance, a student named Charles R. Cross published the paper "On the Focal Length of Microscopic Objectives" in the prestigious *Journal of the Franklin Institute*.[31] Other student publications followed. Indeed, Pickering could happily report that as "more advanced pupils carry on original investigations . . . many unexpected results have been arrived at." Given "the advantages of the laboratory system of teaching physics," he continued, "the tendency of all technical education is now in this direction."[32] Rogers's laboratory-oriented and useful arts philosophy of education was moving the Institute increasingly toward a focus on applied research. As a result of these innovations, the Institute began to acquire a national reputation as a leading center of science and engineering education. As early as 1866, a Harvard friend of Pickering's noted that "the Technological Institute is beginning to make itself heard."[33]

Yet not everything pertaining to the New Education went smoothly. In 1869, four years after classes began, Runkle, Rogers's successor as President, observed that the mechanical engineering department still lacked adequate laboratory facilities for its students. In an effort to strike the right balance between theory and practice, Runkle made arrangements for advanced students to work in the machine shop of the Boston Navy Yard, set up field trips to various local manufacturing facilities, and even organized a special School of Mechanic Arts at MIT, based on a Russian model that sought to introduce students to general shop practices without "overdoing" it and replicating the standard apprentice system found in the industrial sector. Still, as late as 1882, the balance had not been found. Like the faculty in other disciplines, MIT's engineering faculty continued to experiment with different approaches to laboratory practice. Just as technological development itself tended to occur incrementally, so did MIT's approach to engineering education. The problem, as Eliot acknowledged in his 1869 inaugural address as President of Harvard, was "not what to teach, but how to teach"—a sentiment that comported closely with Rogers's view.[34]

In no department was the incremental approach to curriculum development more evident than in the Department of Architecture. The first of its kind in the United States, the department took shape under the direction of Professor William R. Ware, a Harvard-educated Boston architect and structural engineer whose best-known work was Memorial Hall at Harvard. Under Ware's direction, MIT's architecture curriculum underwent continual revision while all the time adhering to Rogers's emphasis on theory and practice. Ware's strength lay on the practical side although he enhanced his program with lectures on art and history, stressing the artistic elements of the discipline. Like so many other architects and engineers of the period, Ware had supplemented his own academic background with hands-on apprentice-style training in design studios, specifically those of Richard Morris Hunt in New York and Edward S. Philbrick in Boston. Upon launching his own architectural partnership with Henry Van Brunt, he developed a similar style of "office teaching" in Boston—an activity that he continued after joining MIT in 1865.[35]

Sensitive to criticisms that too much science might stifle artistic creativity, Ware sought solutions by spending the period between August 1866 and November 1867 in Europe. In London he learned about systematic lecture-based programs of instruction that were sufficiently different from his to provide valuable information about teaching methods that might be tried at MIT. His visits to the Ecole Centrale d'Architecture and the Ecole des Beaux Arts in Paris had an even greater influence on his thinking. The Ecole Centrale proved particularly attractive because its technical orientation and focus on "a thorough grounding in the applied sciences" came close to the theory-practice orientation of MIT. Ware returned from Europe convinced that the French approach to design held the key to mounting a successful program in architecture at MIT.[36]

In fall 1868, Ware introduced a class in design that enrolled four full-time students and twelve "specials." The course grew, and three years later he hired Eugene LeTang, a graduate of the Ecole des Beaux Arts, as a full-time assistant to teach the class. Under LeTang the design-drawing room became a laboratory where students not only learned basic principles but also worked on the development and elaboration of new architectural forms. While Ware continued to identify "architecture as a branch of the Fine Arts," he worked toward building a science-based curriculum that had two primary divisions: construction-practice and composition-design. Of the two, composition-design formed the most distinctive part. At the same time, he encouraged students to move away from the dominant Gothic style of the day toward more functional designs. By the time he departed MIT in 1881 to start a similar program at Columbia University, Ware had developed a complete curriculum that not only enhanced his stature as America's leading architectural educator but also bolstered

FIGURE I.4
The design-drawing room in MIT's Department of Architecture, photographed in 1874. Following the successful pedagogical reforms in chemistry and physics, laboratory-based instruction became central to training MIT students in architecture as well.
Source: Courtesy of the MIT Museum.

MIT's reputation as the place to go for "systematically taught" architectural training.[37]

Ware's approach to education helped to define MIT and advance professional standards not only for architecture but for the emerging professional disciplines of engineering and science as well.[38] The more that science-based education advanced, the more that professionalism advanced, and the more that professionalism advanced, the more that American business and industry turned to modern methods. MIT consequently stood at the center of a fundamental educational shift from craft to professional training in America. Though apparent only to the most astute observers, this shift played a pivotal role in the emergence of the United States as a leading industrial nation and world power.

THE TRIUMPH OF MENS ET MANUS

William Rogers returned to MIT as President pro tempore after John D. Runkle resigned in 1878. Much had transpired during Rogers's absence. From an educational standpoint, it had been a time when the Institute defined itself, becoming well known for its laboratory approach to education, and attained the stature of the country's leading "Scientific School."[39]

Rogers attributed MIT's success to "the inspiration of modern ideas."[40] But modernity was not something that everyone in the world of higher education embraced. Proponents of traditional classical education viewed Rogers and MIT as a hotbed of utilitarianism, a school of practice whose students lacked the finer qualities that cultivated intellectual sophistication and gentlemanly standing. Even within the world of science, Rogers had critics, chief among them Professor Louis Agassiz of Harvard. Rogers and Agassiz crossed swords many times. Agassiz, a zoologist and geologist, stood for established museum-oriented, classificatory traditions in science; Rogers represented a more analytic and experimental approach. A believer in God's direct hand in nature, Agassiz excoriated Darwin's theory of evolution while Rogers defended it. Agassiz advocated a pure science ideal while Rogers promoted a useful arts approach that comported with the new industrial order. The differences between Agassiz and Rogers encapsulated ongoing tensions between so-called "ancients" and "moderns" that began during the antebellum period, and then played out during the rest of the nineteenth century. Ultimately Rogers's vision became the prevailing model of university education in the United States.[41]

Although plagued by chronic financial problems, the Institute nonetheless grew from fifteen students in 1865 to three hundred by 1881. Initially a majority of them were special students who held full-time jobs in local businesses and industrial enterprises while studying part-time at MIT. Eventually the "specials" gave way to full-time four-year students, but it was not until the 1880s that regular students outnumbered special students. In the meantime, President Rogers found a successor for himself in Francis Amasa Walker, a decorated Civil War general and distinguished political economist who came to MIT from Yale after a highly successful stint as director of the U.S. Census of 1880.[42]

Walker brought a new spirit and energy to MIT when he took office in 1882. He used his many contacts and sterling reputation to raise money as well as expand the Institute's overcrowded physical facilities, adding five new buildings to the Back Bay campus between 1883 and 1897. Walker also doubled the size of the faculty and presided over the creation of several new departments, notably electrical engineering in 1882 and chemical engineering in 1888. Both added to MIT's growing reputation. Although the Institute continued to experience annual deficits during the 1880s and 1890s, expansion under Walker continued unabated. In 1881 the Institute enrolled 302 students; by 1891 enrollment topped 1,000 for the first time.[43] In his annual report of 1894, Walker noted that "there is now not a State in the Union without an institution in which more or less of a course in Engineering is laid out. Some of these are classical institutions of long standing and high repute, which are as rapidly as possible transforming to meet the wants of the age."[44] He concluded, "If, indeed, 'imitation is the sincerest form of flattery,' . . . certainly the surviving founders of the Massachusetts Institute of Technology . . . have reason to rejoice that the battle of the New Education is won."[45]

William Rogers would certainly have rejoiced at these words, but sadly, he died at the podium while speaking at General Walker's inauguration. No one would have denied, though, that thanks largely to his forward-thinking educational philosophy and recruitment of a first-rate faculty, the Institute was producing excellent graduates who were playing significant roles in building a modern industrial America. Among them were Arthur D. Little, class of 1885 and founder of the commercial research laboratory that bore his name, and Pierre S. du Pont, class of 1890, president of his family's famous chemical company whose name is often associated with the origins of the modern multidivisional corporation. There were many others. Perhaps the highest compliment came from the leading inventor in the United States, Thomas Alva Edison. When asked by a reporter where he would send his sons to college, Edison unhesitatingly replied, "Yes, to the Massachusetts Institute of Technology."[46]

Emblazoned on MIT's great seal are the Latin words Mens et Manus—Mind and Hand. No one knows exactly who coined the phrase. We do know that the MIT Corporation adopted the seal in December 1864, based on a recommendation of a special committee constituted for the

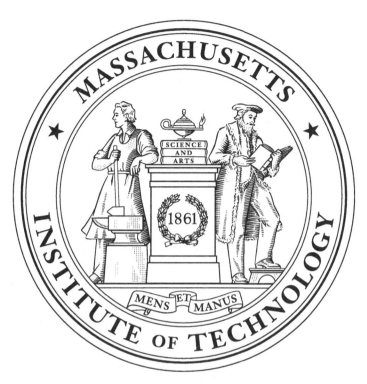

FIGURE 1.5

The MIT seal, featuring the motto "Mens et Manus." The MIT Corporation formally adopted the seal in December 1864.

purpose. William Rogers served on the committee and doubtless played a central role not only in designing the seal (see illustration) but also in devising the motto.[47] The phrase aptly expressed the defining academic values of MIT, and the aim to produce both cutting-edge scientists and engineers.

MIT's success had required individuals with political acumen as well as vision, particularly the ability to cultivate public and private support, fend off critics of an innovative curriculum, and resist numerous attempts to merge MIT with Harvard. Rogers, Runkle, and Walker possessed these attributes. From the beginning, they chose not to insulate the Institute from the outside world but rather to engage it. Mens et Manus—to this day, the words represent the essence of an MIT education and a legacy with which anyone affiliated with the Institute can readily identify.

NOTES

1. William Johnson Walker to Thomas Hopkins Webb, July 2, 1864, quoted in Julius A. Stratton and Loretta H. Mannix, *Mind and Hand: The Birth of MIT* (Cambridge, MA: MIT Press, 2005), 305.

2. Governor John Andrew signed MIT's charter on April 10, 1861. See Stratton and Mannix, *Mind and Hand*, xvii.

3. A. J. Angulo, *William Barton Rogers and the Idea of MIT* (Baltimore: Johns Hopkins University Press, 2009), 120–121.

4. For information on Rogers and his family, see Angulo, *William Barton Rogers*; Stratton and Mannix, *Mind and Hand*; Samuel C. Prescott, *When MIT Was "Boston Tech"* (Cambridge, MA: Technology Press, 1954).

5. For entry to the vast literature on the early Industrial Revolution in the United States, see Brooke Hindle and Steven Lubar, *Engines of Change: The American Industrial Revolution, 1790–1860* (Washington, DC: Smithsonian Institution Press, 1986); Daniel Walker Howe, *What Hath God Wrought: The Transformation of America, 1815–1848* (New York: Oxford University Press, 2007); John F. Kasson, *Civilizing the Machine* (New York: Viking, 1976); Bruce Laurie, *Artisans into Workers: Labor in Nineteenth-Century America* (New York: Noonday Press, 1985); Douglas C. Miller, *The Birth of Modern America, 1820–1860* (Indianapolis: Bobbs-Merrill, 1970); George R. Taylor, *The Transportation Revolution, 1815–1860* (New York: Holt, Rinehart and Winston, 1951). For a historiographical review of the literature on the early Industrial Revolution in the United States (ca. 1790–1865), see Merritt Roe Smith and Robert Martello, "Taking Stock of the Industrial Revolution in America," in *Reconceptualizing the Industrial Revolution*, ed. Jeff Horn, Leonard Rosenband, and Merritt Roe Smith (Cambridge, MA: MIT Press, 2010).

6. For information on Long, McNeill, Whistler, and the B&O Railroad, see Forest C. Hill, *Roads, Rails, and Waterways: The Army Engineers and Early Transportation* (Norman: University of Oklahoma Press, 1957); Charles F. O'Connell, "The Corps of Engineers and the Rise of Modern Management," in *Military Enterprise and Technological Change*, ed. Merritt Roe Smith (Cambridge, MA: MIT Press, 1985), 87–116; Daniel H. Calhoun, *The American Civil Engineer* (Cambridge, MA: MIT Press, 1960).

7. William Barton Rogers, *The Life and Letters of William Barton Rogers*, ed. Emma Savage Rogers (Boston: Houghton Mifflin, 1896), 1:60.

8. Rogers, *Life and Letters*, 1:88–92. See also Prescott, *When MIT Was "Boston Tech,"* 9–10.

9. Robert Rogers to William Barton Rogers, January 7, 1833, in Rogers, *Life and Letters*, 1:101; also quoted in Prescott, *When MIT Was "Boston Tech,"* 10.

10. Stratton and Mannix, *Mind and Hand*, 49.

11. Internal improvement refers to the building of roads, canals, railroads, and various waterway improvements during the early nineteenth century. It was an integral part of Henry Clay's famous "American System" program for economic development after the War of 1812 and remained a much-debated political issue up through the Civil War.

12. On the Virginia survey and its significance, see Sean Patrick Adams, *Old Dominion, Industrial Commonwealth: Coal, Politics, and Economy in Antebellum America* (Baltimore: Johns Hopkins University Press, 2004), 119–151.

13. William Barton Rogers, quoted in ibid., 128.

14. For an excellent discussion of the Virginia and Pennsylvania geologic surveys (1835–1842), see Adams, *Old Dominion*, 119–151. For an equally enlightening discussion of the surveys in the larger scientific context, see Paul Lucier, *Scientists and Swindlers: Consulting on Coal and Oil in America, 1820–1890* (Baltimore: Johns Hopkins University Press, 2008).

15. Rogers, *Life and Letters*, 1:152–153. See also Adams, *Old Dominion*, 142.

16. Rogers, *Life and Letters*, 1:336.

17. Stratton and Mannix, *Mind and Hand*, 85–87; Prescott, *When MIT Was "Boston Tech,"* 17, 117.

18. Stratton and Mannix, *Mind and Hand*, 87–88; Prescott, *When MIT Was "Boston Tech,"* 24–26. The Lowell Institute would later sponsor a well-known night school located at MIT.

19. Rogers, *Life and Letters*, 1:420–421.

20. Prescott, *When MIT Was "Boston Tech,"* 17.

21. The information for this paragraph is drawn primarily from Prescott, *When MIT Was "Boston Tech,"* 27–31.

22. Ibid., 34–35, 38–39.

23. Stratton and Mannix, *Mind and Hand*, 250–272, esp. 265–269; Prescott, *When MIT Was "Boston Tech,"* 132–138; Angulo, *William Barton Rogers*, 117.

24. On West Point, see Theodore J. Crackel, *West Point* (Lawrence: University Press of Kansas, 2002); Peter M. Molloy, "Technical Education and the Young Republic: West Point as America's Ecole Polytechniques, 1802–1833" (PhD diss., Brown University, 1975); Stephen E. Ambrose, *Duty, Honor, Country: A History of West Point* (Baltimore: Johns Hopkins University Press, 1966). On Rensselaer Polytechnic Institute, see Samuel Rezneck, *Education for a Technological Society: A Sesquicentennial History of Rensselaer Polytechnic Institute* (Troy, NY: Rensselaer Polytechnic Institute, 1968); Stratton and Mannix, *Mind and Hand*, 42–45. For information on Harvard's Lawrence Scientific School, see Stratton and Mannix, *Mind and Hand*, 109–138. For details on Yale's Sheffield School, see Stratton and Mannix, *Mind and Hand*, 51–57.

25. Angulo, *William Barton Rogers*, 77–79, 94–96, 118–119, 134, 154–156. Angulo emphasizes laboratory training as an essential feature of Rogers's educational plan, as do Stratton and Mannix in *Mind and Hand*.

26. Angulo, *William Barton Rogers*, 86–89. For a contrasting view of European influences, see Stratton and Mannix, *Mind and Hand*, 61, 434–436.

27. Stratton and Mannix, *Mind and Hand*, 579–580. On Storer, see ibid., 486–487. On Eliot, see ibid., 278–279.

28. Ibid., 581–582; Prescott, *When MIT Was "Boston Tech,"* 66.

29. Stratton and Mannix, *Mind and Hand*, 583.

30. Ibid., 584.

31. Cross would join the MIT physics faculty and go on to found MIT's Department of Electrical Engineering in 1882. See Karl L. Wildes and Nilo A. Lindgren, *A Century of Electrical Engineering and Computer Science at MIT, 1882–1982* (Cambridge, MA: MIT Press, 1985), 16–30.

32. Edward C. Pickering, quoted in Stratton and Mannix, *Mind and Hand*, 585.

33. Ibid., 582. See also Prescott, *When MIT Was "Boston Tech,"* 69.

34. Angulo, *William Barton Rogers*, 144–147; Eliot quotation on 146.

35. John Andrew Chewning, "William Robert Ware and the Beginnings of Architectural Education in the United States, 1861–1881" (PhD diss., MIT, 1986), 25–31. The following paragraphs draw upon Chewning's assessment of Ware's teaching innovations.

36. Ibid., 67.

37. Ibid., 88, 145–148, 165–166, 231.

38. Ibid., 254–255.

39. On MIT's growing reputation, see Stratton and Mannix, *Mind and Hand*, 520; Prescott, *When MIT Was "Boston Tech,"* 117, 129, 136, 146, 150, 157.

40. William Barton Rogers to John D. Runkle, February 1, 1870, in Rogers, *Life and Letters*, 2: 293.

41. On the Rogers-Agassiz relationship, see Angulo, *William Barton Rogers*, 49–56, 93, 111–115; Stratton and Mannix, *Mind and Hand*, 136–137, 258–265, 288–293.

42. Prescott, *When MIT Was "Boston Tech,"* 124, 129. On Walker's background, see ibid., 107–113; see also James Phinney Munroe, *A Life of Francis Amasa Walker* (New York: Henry Holt, 1923).

43. Although students and student life at MIT during the early years are not well documented, it is clear that a small number of women and even fewer black Americans attended the Institute between 1865 and 1900. MIT was an almost exclusively white male domain and would remain so throughout most of the twentieth century.

44. This is doubtless a reference to Harvard College, particularly the efforts of President Eliot to transform the Lawrence Scientific School into an entity more like MIT. Impressed by MIT's approach to science and engineering education, Eliot tried several times to effect a merger of Harvard and MIT. On the merger issue, see chapter 2 in this volume.

45. Francis Amasa Walker, quoted in Prescott, *When MIT Was "Boston Tech,"* 147.

46. Thomas Alva Edison, quoted in "TAE Responds to Reporters' Questions," ca. 1929, folder 30, box 3, Edison Papers, The Henry Ford (formerly The Henry Ford Museum), Dearborn, Michigan.

47. Thanks to Lois Beattie of the MIT Archives for supplying information about the MIT seal.

BRUCE SINCLAIR

It is hard to talk about MIT without including Harvard in the conversation—and that's not simply because they are close neighbors. Their histories
are tangled in strange and interesting ways; in fact, between 1914 and 1917,
the two schools even graduated engineering students with a joint degree.
So it seemed natural from the beginning to imagine the two might be one.
Charles W. Eliot, Harvard's longtime President (1869–1909), was not the
first to propose a merger, but the idea proved so persuasive to him that he
tried it three times. On the MIT side, however, neither founding President
William Barton Rogers nor Francis Amasa Walker, who led the Institute
from 1881 to 1897, ever doubted for a moment that MIT's special kind of
education required its absolute separation from other forms of higher learning. Maintaining that independence proved one of the greatest challenges
the Institute ever faced.[1]

Part of the problem was that Harvard had already started a school osten-
sibly intended to fulfill the same mission as MIT: to train the future design-
ers and managers of technological enterprises. In 1847, scarcely a year after
Rogers put on paper his first ideas for advanced technical education at MIT,
the textile magnate Abbott Lawrence came to Harvard with a parallel plan
as well as the money with which to carry it out. Lawrence was quite plain
about the kind of education he had in mind. There were ample opportunities in America for advanced study in the classical curriculum, he argued,
"but where can we send those who intend to devote themselves to the
practical applications of science?" Answering that question himself, he gave
Harvard the largest gift it had ever received to create a place for young
men "who intend to enter upon an active life as engineers or chemists
or, in general, as men of science applying their attainments to practical

purposes." To avoid any misunderstanding, he stated in the accompanying letter that the money should "be devoted to the acquisition, illustration, and dissemination of the practical sciences forever."[2]

It would be difficult to imagine a more explicit set of directions—or a more complete and instant disregard of them. Harvard ignored Lawrence's wishes, appointed the zoologist Louis Agassiz as a professor in the new school, and justified the act by claiming that the disinterested search for knowledge was a higher form of intellectual activity. As an abiding consequence, although the school was founded in Lawrence's name, the practically oriented education he wanted to promote languished at the university. Even with modest admission standards and simple degree requirements, only 15 percent of the students enrolled in the Lawrence Scientific School from 1851 to 1865 completed the course work necessary for graduation. Meanwhile, MIT enrolled three times as many students by 1869, graduated most of them, and the gap in numbers widened every year. In Boston, at least, it was easy to believe that the vacuum created by the Lawrence School's continued weakness pulled the Institute into existence, and even the governor of Massachusetts said that something should be done to unite the two.

Eliot's role in the ensuing drama is especially complex. Despite his family's long and close connections to Harvard, he was frustrated in his hopes for the professorship in chemistry at the Lawrence School, which went instead to Wolcott Gibbs, another devotee of pure science ideology like Agassiz. So, in 1865, Eliot accepted an offer to become professor of chemistry at the recently established Massachusetts Institute of Technology. He came to the job fresh from a two-year study of French and German technical schools, and a great deal of enthusiasm for the Institute's ambitions as well as a residue of anger from his experience at Harvard—all of which he incorporated into a long, compelling article for the *Atlantic Monthly* in February 1869 titled "The New Education." In the favorite journal of Boston's intelligentsia, Eliot laid out—more comprehensively even than Rogers himself—the argument for separate institutions to train those interested in the practical applications of science. So acute were his observations that they defined for years to come the central issues in the ongoing competition between MIT and the Lawrence Scientific School.

THE RATIONALE FOR SEPARATE TECHNICAL EDUCATION

Eliot aimed to situate technical education in the current economic and social realities of American middle-class life. He thus began his article with an imaginary parent wondering where to send his son for university-level schooling. The father could afford whatever education might be best for the young man, but he was persuaded that practically oriented instruction would prove most appropriate for the "active calling" he expected his son to pursue. Choosing the same kind of language that Lawrence had used when making his substantial gift to Harvard, Eliot linked the creation of wider opportunities in higher education to expanded political freedom and the development of "the prodigious material resources of a vast and new territory." Although engineering, the profession that the parent wanted his son to follow, "did not exist in all the world fifty years ago," Eliot imagined it to be the ideal "American man's life."[3]

He went on to contend that this hypothetical problem was actually one facing every thoughtful parent with sons to educate. Their dilemma was that the country's traditional colleges did not provide training suitable for the "active life," yet it was almost impossible for people to imagine any other form of higher education. In an apparently neutral tone, then, Eliot said his purpose was to describe those American institutions that offered programs of advanced, but practically useful, education.

As he outlined the different forms in which technical education was available in the United States, MIT's professor of chemistry made it clear that only separate institutional arrangements would succeed. Scientific courses of study embedded in the college curriculum, as was the case at Union College, Brown, and the University of Michigan, had this great drawback: the objectives of classical study and technical study were so different that for both to be conducted in the same institution meant each was spoiled. One ought to go to college, Eliot wrote, to study things "for the love of them, without any ulterior objects." By contrast, the technical student had "a practical end constantly in view" no matter how intellectually challenging his studies. It was not that technical studies were improper in a university but rather that the two kinds of instruction could not successfully be taught under the same roof, by the same teachers, although—undoubtedly thinking of the state universities rising up in the West—Eliot

WHIPPLE,   297 Washington St., Boston.

FIGURE 2.1
Charles W. Eliot served as professor of chemistry at MIT between 1865 and 1869, before becoming President of Harvard University. During his brief tenure at MIT, Eliot helped to articulate MIT's mission as an independent institution for technical learning.
Source: Courtesy of the MIT Museum.

allowed it might be done as a temporary expedient "in crude communities where hasty culture is as natural as fast eating."[4]

Nor was practical instruction any better served in those scientific schools allied with established colleges, such as the Sheffield Scientific School at Yale. "Anybody, no matter how ignorant," could enroll in the chemistry department at the Sheffield School, according to Eliot, and that lack of scholarly standards made its students "ugly ducklings." But he directed his harshest criticism of this institutional arrangement at Harvard's Lawrence Scientific School. Administratively it was a law unto itself, and it lacked any "common discipline" of an intellectual kind. Indeed, he pointed out, the opposite was true: a pupil who had studied only chemistry or engineering—someone "densely ignorant" of anything else—could still obtain the school's degree. It was a place, he pronounced, "singularly ill-adapted to the wants of the average American boy of eighteen."[5]

Against all those alternatives, Eliot claimed, the Massachusetts Institute of Technology—the independent scientific school—provided the ideal form for technical education. It offered the most thorough training with a series of four-year courses of study in various fields of applied science, which Eliot described as both "liberal and practical." In discussing MIT, Eliot made two other arguments notable for their subsequent relevance. First, he dismissed the idea that technical study would produce a one-sided person compared with the rounded person of college culture. Echoing Lawrence's view, Eliot claimed that people were born with different talents and that the function of education was precisely to develop innate capabilities. Besides, to think of someone's brain as a sphere that gets filled out equally in every direction was, as he put it, "to be betrayed by metaphor." Furthermore, a properly organized technical education, such as the one offered at MIT, aimed at a rigorous training of reasoning powers, not merely practical exercises.[6]

These observations on the state of higher technical learning in the United States and its centrality to great purposes carried another important conviction. Eliot believed that American educational institutions had "to grow out of the soil"; they had to be the matured fruits of social and political "habits." In an argument reminiscent of Ralph Waldo Emerson's "American Scholar" address, he claimed that technical schools ought to embody entirely national experiences and concerns: "The average American does not eat, drink, sleep, work or amuse himself like an average European. He wants different tools, carriages, cars, steamboats, clothes, medicines, and houses." Men trained to meet those needs had to be schooled in American institutions, and this particular focus on technical education had the largest possible consequence. Political liberty, social mobility, and a wealth of natural resources gave the United States great advantages among the nations of the world, and Eliot imagined a comparably grave responsibility for the country's educational institutions. Out of the kind of idealism that had inspired mechanics' institutes and a host of other Jacksonian-era associations created to make applied science the vehicle of democratic aspirations, he fashioned an educational program for the last half of the nineteenth century with an eye toward the nation's burgeoning industrial future and the concerns of the "enlightened parent."[7]

SLIGHTLY HOSTILE TAKEOVERS: NINETEENTH-CENTURY ATTEMPTS

The "New Education" article was, at one and the same time, a damning indictment of Harvard's Lawrence School, a resounding defense of MIT, and the best argument in the world for keeping them apart. But only a few months later, in a dramatic reversal, the author himself actually began a campaign to absorb MIT into the Lawrence School. This odd turn of events came about because, on the strength of the "New Education" piece, Eliot was chosen as Harvard's new President. (His article's starchy tone probably also explains why he was elected on a split vote.) Fresh from MIT and convinced of the coherence of its curriculum, Eliot proposed a solution to the disorganized drift of the Lawrence School that looked simple and direct: absorb MIT into Lawrence—staff, students, and all.

The notion had an appealing symmetry: on the one hand, MIT attracted large numbers of students, but was continually strapped for money because it had no endowment to speak of; on the other hand, Harvard had considerable funds to support instruction in the application of science, but apparently little talent for providing it. To many leading figures in the community, the obvious solution was to unify all science and engineering instruction at the two schools into a single operation, and fund the entire enterprise with the combined endowments.

That was the essence of an idea Eliot began to discuss informally in fall 1869 with MIT's acting President, John D. Runkle, who had taken over when ill health forced Rogers to retire. There were those who suspected

Eliot's plan masked a grab for power, but most of his former colleagues at MIT were disposed to think well of him at first, especially because of the *Atlantic Monthly* article. As Runkle wrote early in their discussions, "I do hope, my dear Eliot, that some way will be found to secure a co-operation satisfactory to all and unjust to none."[8]

Harvard's wealth was a powerful inducement for MIT. Without some new source of funds, Runkle wrote to Rogers, he did not see how he could keep MIT's best professors from accepting higher-paying positions elsewhere. In a different way, money was also a problem for Harvard: its corporation could not bring itself to relinquish control over the funds for technical education. Harvard's corporation members feared that such a move would signal that the new institution would be a partnership of equals—a price too high for them to bear. Consequently, when the negotiating positions unfolded, it became increasingly apparent to the members of MIT's corporation that unification meant subordination and, sooner or later, the loss of their particular educational vision—the very vision that Eliot had celebrated in his "New Education" essay. R. C. Greenleaf, a member of the MIT negotiating committee, captured the spirit of the deliberations when he acidly reported to Rogers, "I have done all I could to prevent the absorption of the School into the dead carcass of Harvard's School of Science."[9] For his part, Rogers remained sure of the need for MIT's separate, distinctive existence, and the takeover bid reinforced that opinion.

As President of Harvard, Eliot tried two more times to broker a merger: in 1878 because MIT's finances had fallen to a low ebb, and in the 1890s because the Lawrence School's problems remained unsolved. The essential features of all three efforts were not vitally different from each other. Harvard's people tended to feel that MIT ought to yield a little of its independence for so distinguished an association (especially since Eliot offered to name the united school after Rogers). More difficult for the MIT engineers to swallow was the stipulation that technical students ought first to graduate from Harvard College in order to acquire a proper education.[10] The effect of these negotiations was to confirm in the minds of MIT men that their fortunes were far from safe in the hands of the men of elevated culture.

Despite what he had claimed in his "New Education" article, Eliot's fusion schemes always had in the background the idea that engineering might become a professional course of study, such as law or medicine, which one pursued after college. The Lawrence School's administrators continued to flirt with this idea, too, but in fact few postgraduate students showed up. In the meantime, Lawrence's undergraduate enrollments continued to decline—down to fourteen in 1886—as more and more Harvard students took advantage of the opportunity to enroll in Lawrence's scientific courses but take their degrees at the college. As a result, in 1890, when the college and the Lawrence School came under the jurisdiction of Harvard's Faculty of Arts and Sciences, there were serious questions about any further need for the school.

But just one year later, the Lawrence School found a fresh champion when Nathaniel Southgate Shaler took over as dean. Shaler passionately believed that an engineering degree from Harvard ought to have more value than one from MIT. That meant he had to frame important distinctions between the two programs. He did so in an essay published in the *Atlantic Monthly* of August 1893, in which he launched a direct public attack on the kind of schooling MIT provided.

Shaler began with a critique of MIT's institutional form. The fate of independent schools waxed or waned according to their luck in securing effective leadership, he claimed, whereas universities had over the years worked out permanent administrative foundations and were assured of continued stability. Besides, he said, it had come "to be accepted that an American university was incomplete without a school of applied science." He intended to suggest, of course, that this was the "normal" institutional arrangement for technical education in the United States, and he went on to make a case for the university as the only "perfectly successful agent" of instruction ever devised.[11]

Shaler proceeded, with some malice, to compare educational content in the two schools. The independent technical institute originated in the Old World's class rigidities, it was limited in perspective, and its students were narrowly trained. On the one hand, Shaler argued from principle: "The more fit the youth at graduation for the details of a special employment, the less likely he is to have the broad foundation on which his subsequent development must to a great extent depend." On the other hand, he slyly evoked old prejudices, speaking of "trade work" and learning directed at "immediate utility," culturally one-sided technical school graduates trained in "craft sense" for "particular tasks," and an approach to education dominated by "the uncivilized humor of monetary enjoyment." Against those images, Shaler juxtaposed the "truly academic atmosphere of the university, where learning was followed for its own sake."[12]

Much as Shaler depended on these contrasts in educational style to make his point, he knew from MIT's experience that four-year engineering programs attracted students. So he argued that the recent introduction of the elective system at Harvard gave "the earnest, self-guiding young man" the opportunity to fashion a useful course of studies "within the free atmosphere of academic culture."[13] If practical experience were required, it could be had by volunteering to work in the industry of choice during summer vacations.

In this artful fashion, Shaler legitimized, within an ideological context of amateurism, the single-minded pursuit of knowledge for the sake of career. Then with even more audacity, he asserted that in its ability to incorporate applied science training and a liberal arts education under one roof—the very thing that Eliot had said was impossible—Harvard had shown the way to eliminate "prejudices of caste." The American university, offering what Shaler called "culture on a common ground," was to be the mechanism for eliminating those old invidious distinctions between occupations. Thus, engineering would be lifted to the status of the other learned professions through technical education conducted in an atmosphere in which "all the well-trained intellectual service of mankind" was equally valued, and it was this essentially democratic spirit, Shaler claimed, that made universities the "epitome of our culture."[14]

Though he never mentioned MIT by name, no one in Boston could have doubted which independent technical school Shaler had in mind, and MIT's President Walker immediately wrote a hot reply. If universities were so hospitable to enlarged intellectual enterprise, Walker asked, why had the Lawrence School such an unfortunate history? In any event, the crucial point in the organization of technical education was not institutional affiliation, Walker maintained, but prejudice. What really happened, he asked rhetorically, when technical education was "put out to nurse with representatives of classical culture?" To what extent were technical students advantaged by academic atmosphere, by what Shaler called "educative companionship"? Walker's answers rested on his strong sense of MIT's particular academic direction. With a biting image, he spoke of college students "loafing in academic groves," and in a casual, ruminative way "browsing around among the varied foliage and herbage of a great university." He contrasted that aimlessness as well as the frivolity of the college lifestyle with the energy and purpose of technical students. Walker also attacked the social distinctions inherent in Shaler's argument. Why should aspiring engineers prefer a place "where the stained fingers and rough clothes of the laboratory mark them as belonging to a class less distinguished than students of classics or philosophy?"[15]

These contrasts provided nourishment for the Institute's people, but even as the school's increased alumni population began to articulate its own sense of MIT's special significance, the terms of the debate were suddenly and utterly transformed in 1903 by another bequest to Harvard for the purposes of technical education.

This time, and with orders of magnitude more money, the benefactor was the multimillionaire Gordon McKay, a self-made inventor and manufacturer who had considerably increased his wealth through fortunate western mining investments (for which he enjoyed the professional advice of his good friend, geologist, and Cambridge neighbor, Shaler). The money originally came from the purchase of a patent for shoemaking machinery that McKay improved on, and he chose to believe that mechanical ingenuity had created his wealth. That sense of things led him to think of helping young men with a technical flair to acquire the formal education he never had. Shaler's crucial role was to convince McKay that the money should come to Harvard instead of MIT, and he did it with exactly the kind of arguments that he had advanced in his magazine article.[16] Indeed, there is every reason to believe that's why he wrote it in the first place.

No doubt Shaler saw McKay's money as a way to secure the Lawrence School against Eliot's merger schemes. McKay specifically intended the money to support technical training, especially in mechanical engineering, and he wanted the instruction provided at all levels so that public school students might be admitted into Harvard as well as those whose family's wealth had provided better educational advantages. He also meant his bequest to provide handsome salaries for members of the engineering faculty, extensive laboratories, and ample classrooms. But even as Shaler imagined these rich prospects for his school while on a long-delayed sabbatical tour of Europe in 1904, he learned to his dismay that once again Harvard and MIT were considering some form of collaboration.

THE NEAR MISS OF 1905

This time it was not Harvard's Eliot but instead MIT's fifth President, Henry Pritchett, who made the overture, and his reasons for reconsidering an alliance with Harvard are easy to identify. Far and away, the most important factor was McKay's money. MIT's leaders were chilled by the thought of a powerful rival springing up at Harvard—better housed, better equipped, better staffed, and better able to attract the serious-minded, hardworking students whom the technical school had always seen as its particular clientele. Besides, Pritchett knew that the Institute's own classrooms and laboratories were overcrowded, that there were no large lecture rooms at MIT, and that in its Back Bay location in Boston, without a real campus, there were few possibilities for an attractive kind of student life.

In no time, Boston's newspapers were full of the story. The *Boston Record* jumped to the most spectacular conclusion when it reported that MIT's Pritchett was to succeed Eliot as President of Harvard in order to administer both institutions. The article also linked this scenario to large real estate transactions and the completion of the Charles River Dam, then under construction. The *Boston Transcript* offered its readers a simpler but equally intriguing analysis when it claimed the whole idea was "one of Mr. Carnegie's schemes."[17]

There were elements of truth in all these accounts, but as the weeks went by, people started to think about the other implications of such an alliance, and more reasons in favor of it began to surface. One of the most compelling was the threatening educational competition from the state universities of the West. In the East, a region long accustomed to privately supported higher education, the thought of those rapidly expanding public schools "virtually giving away instruction and training" was an awful prospect.[18] The only way that Harvard and MIT could continue to attract students, despite their higher tuition fees and the greater expense of urban living, according to the *Boston Transcript*, was "by keeping ahead of the procession in methods and results," which clearly demanded the efficient pooling of their joint resources.[19] Other facts impressed themselves on the minds of administrators at MIT. Its Boston property had no more space for expansion although every faculty member felt the need for additional classrooms and better laboratories, not to speak of their wishes for improved equipment and larger salaries. But the annual report of MIT's treasurer for

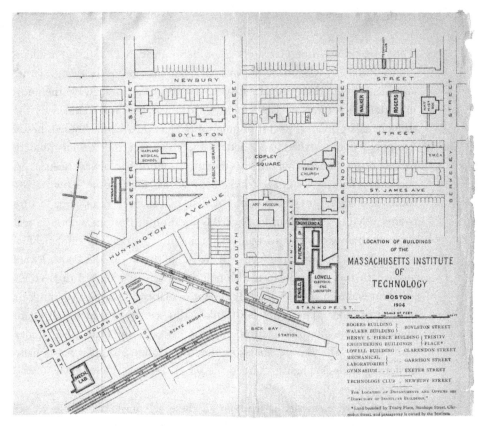

FIGURE 2.2

Map showing MIT's Boston campus in 1904, nestled in the Back Bay region. A lack of room for expansion at the original site provided one motivation for a possible merger between MIT and Harvard.

Source: Courtesy of the MIT Museum.

1903 showed a deficit of $34,000 (nearly $800,000 in 2008 dollars), almost as much as Lawrence had originally given to Harvard. Worse still, an analysis of the school's financial history over the preceding decade revealed that the gap between expenses and income was steadily widening. At the same time, Back Bay real estate had risen substantially in value since the Institute was established, and that suggested moving to a less expensive but larger site to take advantage of the increase.

In addition, there were neither dormitory nor athletic facilities for students at the old stand. In MIT's early years, most of its students lived at home, and because of the school's consciously serious style, no provisions for recreation had been made. But by 1904, when the enrollment neared two thousand, a real demand for housing and the ingredients of campus life had emerged. By that time, too, some of the school's prominent figures had begun to argue the need for a broader curriculum with wider possibilities for general cultural and social exchange. The chief spokesperson for this point of view was John Ripley Freeman, a graduate of the class of 1876, former president of the Alumni Association, and a successful businessperson who maintained an extensive consulting practice in hydraulic engineering.[20] The great industrial changes of the past several years and Freeman's own widening acquaintance with the leaders of those changes led him to see the necessity for something more than technical training if MIT's graduates were to fill the top positions in U.S. corporations. Freeman elaborated his position at length in lectures to alumni groups, and the metaphor he used to make his point was particularly evocative. If MIT stayed as it was while Harvard created extensive programs of pure and applied science, the Institute would end up, he said, educating the corporals of industry instead of its captains. It was not simply the threat of new McKay-financed laboratories that Freeman had in mind; it was Harvard's unquestioned ability to provide its students with a smooth cultural finish.

Freeman's advocacy of broad culture had other foundations. In 1901 he had been appointed chief engineer to the Committee on the Charles River Dam, a group impaneled by the Massachusetts legislature to investigate the possibilities of a dam near the mouth of that river. Ever since Boston's Back Bay had been developed in the mid-nineteenth century, the potentialities of further riverfront improvement, on both banks, were obvious. But many questions (for example, could the waterway carry off the sewage that flowed into it as effectively if dammed) led the state to appoint an investigating committee, chaired by MIT President Pritchett. The committee's report, issued in 1903, encouraged the Boston financier Henry Lee Higginson, in concert with Andrew Carnegie and others, to buy up the land adjacent to Soldier's Field (on the Boston side of the river, opposite Harvard) in the event that MIT might be moved to that site or some other educational use be found for it.

By damming the Charles to create a basin stretching from the river's mouth back upstream for several miles, the ugly mudflats, which were exposed and smelly at low tide, would always be covered. A dam would also eliminate the river's swift and dangerous tidal currents, making it more usable for boating and other recreational purposes.[21] The leap from that idea to the picture of grassy banks and university grounds sloping down to the river, just as at Oxford and Cambridge, captivated Freeman, and it became his great crusade to re-create that picturesque setting along the Charles.

None of Freeman's arguments persuaded either MIT's faculty or other alumni. By May 1904, an aroused group of graduates convened a special meeting to debate the question of "merger" (as its opponents insisted on calling it). Their most widely voiced reason against union was that in any connection with Harvard, MIT would inevitably lose its cherished independence. That was a shorthand way of expressing the fear that after having struggled for years to gain a place in the U.S. educational establishment, technical training would once again be reduced to subordination in that citadel of liberal culture.

FIGURE 2.3

Alumni from MIT's class of 1876 in a parade at the alumni outing in 1904. The message on their banner—"Spirit of '76 Independence"—made clear their disapproval of the proposed merger between MIT and Harvard.
Source: Courtesy of the MIT Museum.

These discussions for and against alliance extended into spring 1905, by which time the administrators of both institutions had worked out what they considered a mutually agreeable plan of cooperation. Its essential features were that MIT would take over all applied science training for both schools, using the income from the Lawrence Scientific School's endowment, three-fifths of the McKay bequest, and MIT's own funds to support the effort. In exchange, MIT would relocate to a property on the other side of the Charles, near Soldier's Field, while retaining its own name and effective control over its own affairs ("home rule," as some called it).

When the plan was presented, however, the opposition's position had hardened considerably. MIT's Association of Class Secretaries reported that its survey showed 95 percent of the school's former students believed MIT should maintain its "absolute independence." The news from an administration-conducted poll was scarcely better: it revealed that alumni were opposed to union by a three-to-one margin—about the same as the faculty vote. Nevertheless, President Pritchett decided that since he had responsibility for the decision, he also had the power to make it, and he pushed through the corporation a twenty-three to fifteen vote in favor of alliance. Incensed alumni wrote indignantly to each other about "such extraordinary exercise of corporate power," and there was dark talk about having Pritchett's head for it.[22] But in the end, it was a court of law, not faculty or alumni opposition, that scuttled the 1905 MIT-Harvard union. One of the conditions of the plan was that MIT realize income from the sale of its Back Bay property, which had come in the form of a restricted bequest, and the court ruled the Institute could not do that. The court's decision also meant that Pritchett could not successfully continue as MIT President, so when he was offered the directorship of the Carnegie Foundation later that year, he did not hesitate in accepting it.

On the Harvard side of the river, President Eliot now had to find another solution to his problems with technical instruction. He returned to the idea of engineering as postgraduate education and replaced the Lawrence School with the Graduate School of Applied Science. The new graduate school offered degree programs in a wide variety of engineering subjects in a curriculum that took seven years of study, at the end of which the student was awarded a bachelor of arts from the college and a master's degree in engineering.[23]

THE TRIUMPH OF SINGULARITY

Though the turmoil that had convulsed MIT for the better part of a year came to an abrupt, almost anticlimactic end, the idea of a combined program in applied science, funded with McKay money, still appealed to the leading figures of both institutions. Shortly after he became the sixth President of MIT, Richard C. Maclaurin wrote to Abbott Lawrence Lowell, the new President of Harvard, that a connection between the two schools seemed "so natural and desireable [sic]."[24] But Maclaurin, a New Zealand–born, Cambridge University–trained physicist who had been lured to MIT from Columbia, understood the complicated dynamics of this interschool rivalry, and from the start, he simultaneously pushed the search for a new location and more money. The idea of a new campus had wide support since the alliance controversy in 1905 had made the need for one apparent. As he looked at various sites in Boston and Cambridge, Maclaurin successfully courted du Pont family alumni for the purchase price. In 1911 he managed to get a $1 million grant from the Massachusetts Commonwealth (roughly $23 million in 2008 dollars), a figure that tripled MIT's endowment. Then in the following year, George Eastman, Kodak's founding genius, but a person known only as "Mr. Smith" to everyone but Maclaurin, offered $2.5 million for buildings on a new site (more than $57 million in 2008 dollars).

In his initial letter to Eastman, President Maclaurin characterized the Institute as a place of small and local beginnings that had steadily grown to national stature and international reputation. He suggested as well that the school's preeminence was due to the distinctive way it had integrated science into its curriculum. When Maclaurin finally met Eastman in person, he revealed his hopes for a cooperative effort with Harvard to create an institution without peer in the world.[25] Maclaurin used all these arguments, including collaboration with Harvard, to fashion a powerful new image for MIT: it was to be "a great *national* school," based on natural science.[26] His demonstrated skill at raising money gave the ideas considerable momentum. In fact, Maclaurin succeeded so well in convincing others of MIT's special yet different destiny that when the negotiations with President Lowell toward an alliance with Harvard were made public in 1912, it finally seemed a union of equals, each institution in its own sphere, and it is difficult to find in the historical records even a murmur of dissent.

Maclaurin recognized the political liability of the Boston site adjacent to Soldier's Field, tied as it was to the failed 1905 plan, and so he decided on the Cambridge location that the school now occupies. In another calculated move, he engaged William Welles Bosworth, John D. Rockefeller's favorite architect, to design the new buildings with clean lines and white facades of Indiana limestone. The neoclassic style bespoke cool rationality and abstract truth, though of a different kind than that distilled in Harvard's old brick buildings. Those who saw the MIT buildings for the first time at the dedication ceremonies in June 1916 were astonished by their appearance: "That stately shrine across the Charles" was one description. "The white and splendid goal" was another.[27] For MIT people, the new buildings were evidence that in the realm of elevated accomplishment, engineering had come of age.

FIGURE 2.4
Festivities to mark the inauguration of MIT's new campus in Cambridge, along the Charles River, in 1916.
Source: Courtesy of the MIT Museum.

To one degree or another, all the events of the three-day celebration to mark the Institute's move across the river centered on the transition from old to new, from past to future. An extravagantly decorated boat, built along the lines of a seventeenth-century Venetian state barge, carried the school's corporate charter and seal, archives, and faculty from the Boston side of the river to the Cambridge shore. That pageant and the subsequent theatrical production, called the "Masque of Power," were the creations of Ralph Adams Cram, an architect steeped in medievalism whose ideas, ironically enough, proved exactly appropriate. The essence of the masque's theme was the direct connection between engineering and noble human attributes. The production itself, however, was far more elaborate, and involved masses of people, highly stylized dances, costumes, music, lighting effects, and conceits such as a steam curtain at the edge of the stage. The culmination of the masque came when architect Cram, playing the role of Merlin, brought the elements of nature to the feet of Alma Mater, the spirit of MIT, to be held in trust by her for the welfare of future generations.[28]

It was not the moral tone that impressed people but rather the aesthetic plane of the production, which like the new buildings, seemed another proof of maturation. To the engineers, the pageants provided "the perfect symbol of what has been happening all these years. The necessity of concentrating on the purely utilitarian has passed; the chrysalis has lived out its time and the butterfly of art has crept out to try its wings."[29]

While these engineers described a world in which their work provided the basis, indeed was the precondition, for a more exalted intellectual life, MIT's physicist President, Maclaurin, took the opposite stand. He swept away the distinctions between applied and pure science, and then, in a model that was to become increasingly familiar, made the first dependent on the second. Indeed, Maclaurin's reformulation of the Institute's goals and his talent for raising money so energized the school that in 1917, when the court once again ruled against union with Harvard (this time on the grounds that it violated the terms of McKay's will), the decision hardly changed anything. The sense of having finally arrived, of being able to control their own destiny, made it comfortable for MIT people to live with the court's decision and without McKay's money. Future collaboration with Harvard would be of the organic kind that both Rogers and Eliot had

FIGURE 2.5

Ceremonial transfer of MIT's Boston campus to its new Cambridge location, 1916. The ornate boat—designed in imitation of a seventeenth-century Venetian state barge—ferried MIT's faculty and official documents across the Charles River. *Source*: Courtesy of the MIT Museum.

imagined: cooperative efforts that grew out of particular circumstances and reinforced the basic character of each institution.

NOTES

1. This chapter is substantially based on two previously published works: Bruce Sinclair, "Inventing a Genteel Tradition: MIT Crosses the River," in *New Perspectives on Technology and American Culture*, ed. Bruce Sinclair (Philadelphia: American Philosophical Society, 1986), 1–18; Bruce Sinclair, "Harvard, MIT, and the Ideal Technical Education," in *Science at Harvard University: Historical Perspectives*, ed. Clark A. Elliott and Margaret Rossiter (Bethlehem, PA: Lehigh University Press, 1992), 76–95.

2. Abbott Lawrence to Harvard treasurer Samuel Eliot, June 7, 1847, Corporation Records, Harvard University Archives, Pusey Library, Cambridge, Massachusetts.

3. Charles W. Eliot, "The New Education, I," *Atlantic Monthly* 23 (February 1869): 203–221, on 210–211.

4. Ibid.

5. Ibid., 203.

6. Ibid., 218.

7. Charles W. Eliot, "The New Education, II," *Atlantic Monthly* 23 (March 1869): 358–367, on 365.

8. John D. Runkle to Charles W. Eliot, February 2, 1870, MIT Institute Archives and Special Collections, Cambridge, Massachusetts.

9. R. C. Greenleaf to William Barton Rogers, Boston, July 28, 1870, MIT Institute Archives and Special Collections, Cambridge, Massachusetts.

10. John D. Runkle to William Barton Rogers, January 27, 1870, Rogers Papers, MIT Institute Archives and Special Collections, Cambridge, Massachusetts.

11. Nathaniel Southgate Shaler, "Relations of Academic and Technical Instruction," *Atlantic Monthly* 72 (August 1893): 259–268, on 261.

12. Ibid., 267.

13. Ibid., 265.

14. Ibid., 268.

15. Francis Amasa Walker, "The Technical School and the University," *Atlantic Monthly* 72 (September 1893): 390–395.

16. Nathaniel Southgate Shaler, "Gordon McKay," *Harvard Graduate's Magazine* (June 1905): 569–575.

17. *Boston Record*, January 27, 1904; *Boston Transcript*, January 23, 1904.

18. *Boston Record*, January 27, 1904.

19. "The Proposed Harvard-Technology Merger," *Technology Review* 6 (April 1904): 184; *Boston Transcript*, January 3, 1904 and January 25, 1904.

20. For an extensive obituary, see Vannevar Bush, "John Ripley Freeman, 1855–1932," *Biographical Memoirs of the National Academy of Sciences* 17 (1935): 171–187.

21. For a description of several previous attempts to dam the Charles River, see *Report of the Committee on Charles River Dam* (Boston: Wright and Potter, 1903).

22. Association of Class Secretaries of the Massachusetts Institute of Technology, *Report of Eighth Annual Meeting*, November 15, 1904; E. C. Hultman to various MIT alumni, August 1, 1905, Freeman Papers, MIT Institute Archives and Special Collections, Cambridge, Massachusetts.

23. Hector James Hughes, "Engineering and Other Applied Sciences in the Harvard Engineering School and Its Predecessors," in *The Development of Harvard University since the Inauguration of President Eliot, 1869–1929*, ed. Samuel Eliot Morison (Cambridge, MA: Harvard University Press, 1930), 432.

24. Richard C. Maclaurin to Abbott Lawrence Lowell, June 17, 1909, Lowell Papers, Harvard University Archives, Cambridge, Massachusetts.

25. Richard C. Maclaurin to Mr. Eastman, February 29, 1912, MIT Institute Archives and Special Collections, Cambridge, Massachusetts.

26. Richard C. Maclaurin, "President's Report, December 1916," MIT Institute Archives and Special Collections, Cambridge, Massachusetts.

27. *Technology Review* 18 (July 1916): 468, 479.

28. A full account of the entire celebration appears in *Technology Review* 18 (1916).

29. Ibid., 466.

CHRISTOPHE LÉCUYER

The Massachusetts Institute of Technology experienced a remarkable and thoroughgoing transformation between the early 1910s and the late 1930s. Having operated as an undergraduate engineering school since its formation in 1861, the Institute turned itself into a full-fledged research university. This metamorphosis is all the more noteworthy since rare were the engineering schools that experienced such a complete transformation. In fact, in the period under consideration in this chapter, only the Throop College of Technology, a polytechnic institution offering vocational education in Southern California, made the change: it became the research university known as the California Institute of Technology. Most other undergraduate engineering schools that became research universities did so after World War II, often as late as the 1970s and 1980s.[1]

In tandem with MIT's conversion to research and graduate education, the Institute established increasingly close ties with industrial firms. Faculty members developed scores of research collaborations, and in 1919 MIT administrators launched the Technology Plan (or Tech Plan), which aimed to create even easier back-and-forth relations with industrial patrons. The most powerful of these patrons evolved a clear vision of what they wanted for MIT's future; they also had the resources to make it happen. But from start to finish, the course of MIT's remaking turned out to be long, circuitous, and conflictual.

THREE GROUPS, THREE VISIONS

As the twentieth century got under way, MIT was an undergraduate engineering school with a strong practical bent. Its primary goal was to train engineers for practical vocations in industry. The Institute offered a mix of science courses, practice-oriented engineering courses, and training in specific technical skills. Students took a year of calculus and two years of physics, but most of the curriculum consisted of descriptive courses that emphasized engineering practice and training in a vast array of technical operations. For example, instruction in metallurgy and mining engineering was based on lecture courses that covered metallurgical processes in minute detail and offered an intricate classification of American mining machinery. These lectures were supplemented by work in the teaching laboratory for metallurgy and mining engineering, where students operated machinery and carried out entire processes, from the treatment of ores to weighing, assaying, and analyzing metallurgical products. It was only "by carrying a process to the end," as a faculty member wrote in 1895, that "students could learn how to think metallurgically." A similar approach to engineering education was followed in the Institute's other courses of study, such as mechanical engineering, electrical engineering, and industrial chemistry.[2]

This system of engineering education required the teaching staff to keep abreast of professional practice. Faculty members organized visits to factories and summer schools where students worked in industrial establishments. The staff also consulted for industries in the Boston area, and MIT's administrators appointed practicing engineers as lecturers. In 1901, for example, forty practicing engineers taught courses at MIT in which they discussed their respective industries' latest technical developments. These courses ranged from telephone engineering taught by an engineer from AT&T to cellulose chemistry offered by Arthur D. Little, an MIT graduate and consulting chemist who cofounded what later became a major consulting firm.[3]

In the first decade of the new century, however, groups of young faculty members and administrators increasingly questioned MIT's practical approach to engineering education along with its focus on undergraduate teaching. Many of these faculty members had done their graduate work in Germany, where they had participated in active research programs and witnessed close collaborations between academic laboratories and industrial firms. They were also worried by the growth of engineering schools at midwestern universities as well as the emergence of research universities that increasingly competed with MIT for students and faculty. To confront the rising competition, they sought to introduce a stronger element of

FIGURE 3.1
MIT mechanical engineering students analyze the operation of a Harris-Corliss engine, ca. 1900. Within the decade, prominent voices at MIT began to argue that MIT's curriculum had become too narrowly focused around learning to operate machinery at the expense of training in the sciences.
Source: Courtesy of the MIT Museum.

science in the engineering curricula, and develop research and graduate programs. They also favored closer links with industry. They were not a unified group, though: they held different views of engineering, they gave different meanings to the notion of "industrial service," and they advocated divergent projects for the Institute.

One of the reforming groups was led by Arthur Noyes, a professor of physical chemistry who had received his PhD at the University of Leipzig in 1890. After the failed merger with Harvard prompted the resignation of MIT President Henry Pritchett, Noyes served as acting President for two years (1907–1909). Noyes and his group wanted to transform MIT into a research university focused on the sciences. They thought of engineering as laboratory science and advocated a major overhaul of the curriculum. Engineering education, they contended, should rest on work in the humanities, "scientific investigations of technical problems," and "thorough training in the principles of fundamental sciences and in the scientific method." Only through such reforms could MIT produce a qualitatively different kind of engineer—one who would be a "leader on the scientific side of the development of the industries in the country."[4]

Their concept required building research laboratories and graduate programs in the sciences. The Research Laboratory of Physical Chemistry, which Noyes established in 1903 and then directed, soon became the premier center for research in the field in the United States. In addition to fundamental studies on dilute aqueous solutions, the lab staff developed advanced technologies such as mercury arc rectifiers, which were further refined in industry. Many physical chemists who worked there went on to research careers in industry. The laboratory also graduated MIT's first three PhDs in 1907.[5]

Another important component of Noyes's project for the Institute was forging close links with industry, especially science-based industry. "All those industrial, commercial, and transportation interests which are dependent in any measure on scientific knowledge and investigation," wrote Noyes in his presidential report in 1908, "should be made to feel that the Institute stands ready to place at their service, for the study of their problems, the expert advice of its staff and its laboratory facilities." Noyes was particularly interested in building ties with the industrial research laboratories recently established by DuPont and General Electric, and training researchers for these laboratories.[6]

Competing with Noyes and his allies was a group of faculty members around William Walker, a chemical engineer, and Dugald Jackson, chair of the Department of Electrical Engineering. They wanted to transform MIT into an elite technological school closely connected to industrial pursuits. Their ideas about reforming engineering education at MIT accorded with their view of engineering as management. They thought of engineers as corporate leaders and wanted to train the future managers of U.S. industry. They also believed that technological schools should contribute to America's transformation into a great industrial nation. Only through "a closer alliance between the scientific worker and the actual agencies of production," Walker wrote in 1911, could the country "find its industrial salvation, improve the often mean and sordid living conditions of the working classes, and provide a general moral and spiritual uplift to the great masses of the community."[7] But even Walker and Jackson were not in complete agreement. Walker wanted to serve small and midsize companies, and transform them into competitive concerns through the performance of research and the aggressive introduction of science into their activities. For Jackson technical schools were to serve large firms, such as utilities and large electrical manufacturers.[8]

In order to turn out future corporate managers, Jackson reformed the electrical engineering curricula while Walker transformed MIT's course in industrial chemistry into a chemical engineering program. In 1908, Walker also established the Research Laboratory of Applied Chemistry (RLAC), an engineering laboratory in close contact with industry. In Walker's mind, the RLAC was to be at the center of a national network of technical laboratories placed "at points of best study and attack for a particular industry, such as Gary for iron and steel or in the Lehigh Valley for cement."[9] Unlike Noyes's Research Laboratory of Physical Chemistry, Walker's laboratory remained relatively small through the first half of the 1910s. Attracting modest industrial attention, the RLAC employed only a handful of chemists, tackled problems of uneven scientific and technological value, and ultimately failed to provide the foundation for an MIT-centered network of industrial laboratories.[10]

Competing with the reforming impulses of Noyes, Walker, and Jackson was an influential third group of faculty members around Henry Talbot, chair of the chemistry department, and mechanical engineer Edward Miller. Their group was pragmatic and wanted to modernize MIT without breaking with its tradition of preparing undergraduate students for positions of immediate usefulness in industry. They embraced elements of the reform programs of both Walker and Noyes, such as the establishment of research laboratories and the institutionalization of relations with industrial corporations. But the differences mattered. For example, Walker (as described above) saw research laboratories as a tool for transforming small industrial corporations and making U.S. industry more competitive. For Talbot, however, research laboratories were mainly a more efficient way of achieving what MIT faculty members had been doing since the 1880s: keeping in touch with engineering practice and finding jobs for graduates.[11]

IN THE SERVICE OF INDUSTRY

Tipping the balance in favor of Walker and Jackson were the appointment of Richard Maclaurin as MIT President and his great success in enlisting industrial patronage. Maclaurin, a hard-driving New Zealander, was a mathematical physicist with a doctorate from Cambridge. Soon after his appointment in 1909, he embarked on a program of institutional advancement that gave a central place to industrial service. He wanted to build what he called the "New Technology," a national "school of science that concerned itself with practical affairs" and would "train young men to apply the methods of science to the industrial development of the land." His goal was also to "strengthen the Institute as an instrument of research regarding the problems of industry." In order to transform MIT, Maclaurin enlisted industrial patronage. He raised monies from the du Pont family and George Eastman, the founder of Eastman Kodak. Eastman, who had relied heavily on MIT graduates for the supervision of his company's most complex manufacturing processes, had a high opinion of the school's faculty and student body. He gave the Institute $2.5 million in 1912 (nearly $57 million in 2008 dollars). This was the first of many gifts, which totaled $6 million by 1917 (more than $100 million in 2008 dollars). The gifts supported the construction of the new campus in Cambridge. They also financed the educational and research ventures championed by Walker, Jackson, and like-minded faculty members.[12]

In the mid-1910s, Walker and Jackson organized new master's programs in electrical and chemical engineering, both of which aimed at exposing students to the best engineering practice. For example, Jackson established a cooperative teaching program with General Electric in which students alternated periods of work at General Electric with course work at MIT. Walker's new venture was the School of Chemical Engineering Practice, founded in 1916. The school was composed of five "stations" located at paper, cement, dyestuffs, and electrochemical plants. It offered fourth- and fifth-year chemical engineering students an opportunity to live with and analyze chemical processes conducted on a large scale. Realizing Walker's original goals for the Research Laboratory of Applied Chemistry, the School of Chemical Engineering Practice also performed research for small firms. In residence at each station was a member of MIT's teaching staff who instructed students and at the same time conducted research of interest to the host company. Each station director was salaried by MIT, but the company covered research expenses. Importantly, and controversially, the research findings belonged to the company.[13]

Maclaurin's focus on industrial service increased after World War I. To make up for lost revenues in the immediate aftermath of the war, Maclaurin launched a new fund-raising campaign in 1919. The campaign failed to garner much financial support, though, until Eastman offered $4 million in June 1919 (roughly $50 million in 2008 dollars) on the condition that a matching amount would be raised before the end of the year. Unable to find an adequate source of revenue through traditional fund-raising channels, Maclaurin turned to corporate patronage. With Walker's help, he organized the Tech Plan in fall 1919, a scheme whereby corporations paid MIT an annual retaining fee, and in return gained access to libraries and alumni records while also receiving technical services from members of the teaching staff.[14]

Although the Tech Plan was hastily conceived, it can be seen as the acme of Maclaurin's campaign for industrial research. Maclaurin and Walker built on the relations that the faculty had already developed with industrial corporations. They also relied on their own experience in enlisting corporate patrons for cooperative educational programs and research laboratories. But the Tech Plan also constituted a significant departure from previous practice. Prior collaborative arrangements with industry had been carried out at the laboratory or department level. Instead, the Tech Plan was administered by a central office at MIT, the Division of Industrial Cooperation and Research (DICR), headed by Walker. The other novelty of the Tech Plan was its size. Never had MIT's administrators tried to enlist corporate cooperation on such a large scale. Between October 1919 and January 1920, no less than two hundred corporations subscribed to the plan, bringing in nearly a half-million dollars during its first year of operation (more than $5.3 million in 2008 dollars).[15]

The contractors were extremely diverse in geography, size, and activity. While a majority of the corporations were based in New England and the mid-Atlantic states, there were also a significant number of midwestern companies. Participants included AT&T, major electrical manufacturers such as General Electric, and large steel and rubber corporations, but there were numerous smaller concerns in the textile, paper, and machine-tool industries as well. Corporate motivations for joining the plan seem to have

FIGURE 3.2

MIT students test motors at a General Electric facility in 1920, as part of the MIT–General Electric Cooperative Course in Electrical Engineering.

Source: Courtesy of the MIT Museum.

FIGURE 3.3
Richard C. Maclaurin served as MIT President between 1909 and 1920. He launched the Tech Plan in 1919 in an effort to further solidify MIT's relationships with industrial patrons.
Source: Courtesy of the MIT Museum.

been diverse: the larger firms may have wanted to ensure an adequate supply of engineers and gain a first call on MIT graduates; some of the smaller corporations thought of the Tech Plan as their "scientific insurance" and considered the Institute their "industrial laboratory." In their view, the Tech Plan would enable them to tap into MIT's technical expertise in case of unexpected difficulties in product development or manufacturing.[16]

The Tech Plan intensified tensions between Noyes, who wanted to educate scientists and research engineers, and the group around Walker. Noyes staunchly opposed the Tech Plan, claiming that it would "stifle research in pure science," "throw all of the activities of the Institute into applied research," "commercialize Technology," and make MIT "subservient to a group of corporations."[17] Noyes resigned in November 1919 and moved to the Throop College of Technology, where he played a leading role in its transformation into Caltech, a major science-centered university. Losing its leader and most articulate spokesperson, the group advocating the transformation of MIT into a research university was seriously weakened. Comprising a handful of faculty members in the physics and chemistry departments, the group became a relatively minor force at MIT during most of the 1920s.[18]

When the Tech Plan was still in an early stage of development, President Maclaurin suddenly died. The direction of the Institute was given to an administrative committee chaired by Walker. Its other members were Talbot and Miller, leaders of the third group of faculty members, those who wanted to modernize MIT while keeping its tradition of preparing undergraduate students for positions of immediate usefulness in industry. Walker attempted to use the Tech Plan and the DICR as tools in his crusade for a more competitive and scientific industry. "The Technology Plan," Walker wrote in *Science*, "is a more effective means of introducing technical research to the manufacturer; of making the application of science to industrial problems popular; of creating an appreciation on the part of the leaders of industry of the value of science."[19] To promote industrial research and ensure contract renewals, Walker advocated a policy of aggressively soliciting opportunities to serve contractors with industry visits to the Institute and problem-seeking trips to manufacturing plants. Walker also demanded the power of allocating industry problems to the faculty and creating "a bee-hive of workers, from which men could be detailed to assist a member of the faculty in the solution of problems as the Division obtains them."[20]

Miller and Talbot vigorously fought Walker's goals for the Tech Plan, which they argued would have transformed MIT into an industrial research institute. They forbade Walker to solicit new problems from contractors and denied him the authority to assign problems to faculty members. They then forced him to resign from the administrative committee and the leadership of the DICR. Walker soon left the faculty to head the patent department of Dewey & Almy, a start-up chemical firm based in Cambridge. Miller and Talbot nominated Charles Norton, a physicist known for his development of asbestos-based fire retardant materials, to direct the DICR. In contrast to Walker, who wanted to harness MIT's resources through the Tech Plan to make U.S. industry more "scientific,"

Miller and Talbot wanted the Tech Plan to serve the Institute's interests and enhance its service to industrial companies, especially small corporations. They gave the DICR four objectives: "augment salaries that enable the Institute to keep its men," "supply additional opportunities for research to the teaching staff," help the faculty keep in touch with industrial practice, and serve small firms that could not afford research laboratories. Under the new arrangement, the DICR kept the records of the alumni, arranged meetings between faculty members and industrial corporations, and coordinated faculty research sponsored by industry. These services were also extended to corporations that did not subscribe to the Tech Plan.[21]

Under the leadership of Talbot and Miller (and later, the presidency of Samuel Stratton), industrial service became central to MIT's institutional culture. Research support grew from $56,452 in 1920 to $264,797 in 1927 (or from roughly $700,000 to $3.3 million in 2008 dollars). Norton estimated that more than a third of the teaching staff were actively engaged in research, testing, and commercial analyses for industry by 1929. Similarly, faculty consulting grew considerably. Most engineering department chairs directed consulting firms in downtown Boston. More than half of the MIT staff regularly consulted for outside concerns during the 1920s. The unstated rules that governed the Institute also changed. After the war, faculty appointments and nominations to department chairmanships increasingly depended on consulting and development work. For example, Norton not only headed the DICR but he also became the chair of the Department of Physics in 1922.[22]

These developments led to the partial control of MIT research activities by industrial corporations. Most research contracts gave firms the right to block the publication of results. The RLAC, the largest performer of sponsored research in the 1920s, saw its freedom of action severely curtailed by business patrons. Most of the corporations forbade the RLAC to publish the results of the studies they had commissioned. As it became increasingly clear to faculty members that industrial patronage had significant downsides, Norton reported to MIT's leadership in 1924 that he "was permitted to say very little about the nature and scope of the development work done by our staff through the DICR," owing to the confidential nature of the research projects.[23]

FROM TECHNICAL SCHOOL TO RESEARCH UNIVERSITY

Starting in the late 1920s, the vision of MIT as a research university experienced a resurgence. The impulse for reform came from outside the faculty
this time, from two powerful industrialists: Gerard Swope, president of
General Electric, and Frank Jewett, head of the Bell Telephone Laboratories. (Swope headed the executive committee of the MIT Corporation, and
Jewett was on the advisory committee of the Department of Electrical
Engineering.) With the exception of a few departments (such as chemical
engineering), undergraduate education at MIT had remained oriented
toward practice in the 1920s. In the view of Swope and Jewett, however,
industry no longer needed the practice-oriented engineers whom MIT had
been training since the 1880s. They wanted the Institute to educate engineers with a solid understanding of the sciences—engineers who would be
able to contribute creatively to science-based technologies and industries.
At first Swope and Jewett promoted reforms in the Department of Electrical
Engineering. They advocated a stronger emphasis on science in both the
undergraduate and graduate curricula, and were instrumental in the formation of the honors program. They also pushed for an expanded physics
research program with the paradoxical result that the more active physicists
at the Institute could be found in electrical engineering by the late
1920s.[24]

After their success in the Department of Electrical Engineering, Swope
and Jewett extended the reforms to the Institute as a whole. Swope recruited
Karl Compton to the presidency in 1930. Compton, a well-known experimental physicist and a member of the National Academy of Sciences, was
chair of the Department of Physics at Princeton and had worked as a consultant for General Electric. Swope and Jewett gave the like-minded
Compton the mandate of "introducing a much more powerful element of
fundamental science" into the engineering curricula at MIT. They also
suggested that this reform of the curricula would be best implemented by
strengthening the science departments and appointing "research men" in
engineering.[25]

To implement Swope and Jewett's mandate, Compton could count on
a talented group of faculty and administrators who embraced the research
university vision. They included Vannevar Bush, one of MIT's first PhDs
in engineering, a pioneer in analog computing, and head of the electrical

engineering research laboratory, and Frederick Keyes, the chair of the chemistry department who had worked as a postdoctoral fellow under Noyes, and had later established research laboratories in organic and inorganic chemistry at MIT. Compton financed the curricular reforms and the creation of strong science departments and research programs with funds donated by Eastman. Compton also diverted resources from engineering to science; notably, he eliminated costly degree programs, such as gas and fuel engineering, and raised monies from the Rockefeller Foundation.[26]

President Compton reformed the curricula in the 1930s along the lines that Noyes had advocated two decades earlier. The Institute's goal, as Compton explained in his report for the academic year 1930–1931, was to offer a "technological education," by which he meant an "education in the fundamental principles" along with "a training in their application to important basic processes and problems," rather than a mere "technical education." The ultimate goal was to "produce leaders who would be able to handle the big and difficult problems of organization, production, and development." To achieve these objectives, he reorganized the undergraduate curriculum. Students' first two years were devoted to basic training in mathematics, physics, chemistry, English, and history; the upper-class years offered the opportunity for specialization. This new organization of studies had considerable repercussions on course offerings for first-year students and sophomores. The Department of Chemistry, for instance, had to design a new integrated introductory course in lieu of the traditional ones addressed to engineers and chemists. The reformulation led to considerable infighting, which finally resulted in a course much closer to the older one for chemists than the one for engineers.[27]

Compton also drastically transformed the Department of Physics. He eased out a number of aging faculty members and replaced Norton, a specialist in industrial physics, with John Clarke Slater, a theoretical physicist, as department chair. Compton also transferred physicists from electrical engineering to the Department of Physics, and hired promising Princeton PhDs and national research fellows, such as Robert Van de Graaff, as research associates and junior faculty members. In the mid- to late 1930s, Compton extended his department building efforts to meteorology and biology. In tandem with the appointment of star scientists to the MIT faculty, he built new research facilities. Using funds donated by Eastman, he erected the George Eastman Laboratories which housed the Institute's

FIGURE 3.4

MIT President Karl T. Compton (right) with MIT Vice President Vannevar Bush
in the 1930s. Compton and Bush launched reforms at MIT during the 1930s to
emphasize basic science in research and teaching, and restore a greater measure of
autonomy for MIT in relation to its industrial patrons.

Source: Courtesy of the MIT Museum.

research groups in physics and chemistry. Another important center of research was the Round Hill Laboratory, located in South Dartmouth, Massachusetts, where MIT faculty members had experimented with radio systems since the mid-1920s. At the Round Hill Laboratory, Van de Graaff designed and constructed ever-more-powerful electrostatic accelerators for nuclear physics research.[28]

Compton and Bush (who became Vice President and Dean of Engineering in 1932) pursued similar policies for the engineering departments. They appointed prominent research-oriented engineers to key positions. For example, Jerome Hunsaker, who had received one of MIT's first engineering doctorates and had headed aircraft design for the U.S. Navy, was

FIGURE 3.5
MIT's Building 6, housing the George Eastman Laboratories for chemistry and physics, under construction in 1931. A large gift donated a few years earlier by photography inventor and entrepreneur George Eastman enabled MIT President Karl Compton to enact many of his basic-science reforms.
Source: Courtesy of the MIT Museum.

appointed chair of the Department of Mechanical Engineering. To rejuve-
nate the department, Hunsaker hired several physicists. In metallurgy as
well, MIT administrators attracted two more physicists. Compton appointed
individuals with backgrounds in applied science in the electrical and chemical engineering departments, both of which had already converted to
research. Along with establishing the Eastman Laboratories, these appointments were intended to dramatize the renewed emphasis on science at
MIT.[29]

In a manner reminiscent of Noyes's approach to industrial relations,
Compton and Bush supported close ties with industrial corporations. They
considered service to industry central to the Institute's mission, but they
wanted industrial collaborations to advance, rather than hinder, the Institute's other main missions: teaching and research. These considerations led
Compton and Bush to take command of MIT's relations with industrial
corporations. They were greatly aided in this effort by the onset of the
Great Depression, which led to a significant decrease in industrial research
contracts and weakened the power base of Walker's followers. In 1934,
Compton shut down the ailing Research Laboratory of Applied Chemistry,
which had dealt independently with industrial firms since its founding by
Walker and had allowed business patrons to limit its freedom of action
severely. Compton also charged the DICR, originally formed to administer
the Tech Plan, with controlling the terms of research contracts with industry. After 1932, every research agreement had to be "approved by the
director of the division, acting under the direction of the President," who
made sure that the contract gave publication rights to the research staff and
respected the autonomy of the Institute.[30]

In order to support the Institute's growing research programs, Compton
and Bush also expanded the network of MIT's patrons to include the federal
government. As chair of Franklin Roosevelt's Science Advisory Board from
1933 to 1935, Compton advocated massive federal support to academic
science. In 1936 and 1937, he helped Lyman Briggs, the director of the
National Bureau of Standards, to promote a bill in Congress authorizing
the bureau to give research grants to nonprofit institutions. These efforts
were unsuccessful, however. Key members of the Roosevelt administration
and part of the U.S. science establishment actively opposed greater federal
support to university research. This opposition ultimately doomed Compton's plan for the National Research Administration, a new agency that

would finance fellowships and research grants for nonprofit institutions. Parallel to these unsuccessful attempts, Compton tirelessly solicited government aid for MIT's research programs. From 1932 to 1935, MIT tried to negotiate a contract with the Tennessee Valley Authority for the development of a new system of high-voltage power transmission. A few years later, Compton approached the newly founded National Institutes of Health with a list of proposals. Both of these efforts failed. Nonetheless, MIT did win an increasing number of federal research grants from the Civil Aeronautics Administration, the military, and the Department of Commerce in the late 1930s. Federal support for research at MIT nearly doubled from $23,000 in 1936 to $44,000 in 1940 (from $360,000 to $680,000 in 2008 dollars).[31]

The reforms that President Compton implemented in the 1930s, at the urging of General Electric's Swope and Bell Labs' Jewett, reoriented the Institute: MIT became a full-status research university. Along with the California Institute of Technology, it joined the Association of American Universities, the organization connecting elite research universities, in 1934. The Institute saw its science departments rise to prominence. It also gained a larger degree of autonomy vis-à-vis industry. Because of its new strength in research, MIT even emerged as a rival to General Electric in the late 1930s. Van de Graaff and his team had installed a million-volt X-ray generator at the Huntington Memorial Hospital for cancer treatment. This provoked longtime patron Swope to complain to Compton in 1938 that "MIT was engaging in improper commercial competition with GE" and its medical instrumentation business.[32]

By the late 1930s, MIT was a different institution from the one it had been four decades earlier. From a polytechnic institution training practical engineers for positions of immediate usefulness in industry, it had become a full-fledged research university with leading research and graduate programs in physics, chemistry, electrical engineering, and chemical engineering. Its undergraduate curriculum included significantly more training in the physical sciences, and was aimed at producing engineers and managers who would creatively contribute to U.S. industry. MIT faculty conducted substantially more research for industrial companies than they had performed at the beginning of the century. They also trained a growing number of scientists for employment in corporate research laboratories. Perhaps more important, administrators and faculty members had gained expertise on how to manage relations with industry. They had learned how to collaborate with industrial firms without compromising either the Institute's freedom of action or its educational and research missions.

NOTES

1. This chapter is adapted from Christophe Lécuyer, "The Making of a Science-Based Technological University: Karl Compton, James Killian, and the Reform of MIT, 1930–1957," *Historical Studies in the Physical and Biological Sciences* 23 (1992): 153–180; Christophe Lécuyer, "MIT, Progressive Reform, and 'Industrial Service,' 1890–1920," *Historical Studies in the Physical and Biological Sciences* 26 (1995): 1–54. For other treatments of MIT in the period under consideration in this chapter, see Christophe Lécuyer, "Academic Science and Technology in the Service of Industry: MIT Creates a 'Permeable' Engineering School," *American Economic Review* 88 (1998): 28–33; David Noble, *America by Design* (New York: Knopf, 1977); John W. Servos, "The Industrial Relations of Science: Chemical Engineering at MIT, 1900–1939," *Isis* 81 (1980): 531–549; W. Bernard Carlson, "Academic Entrepreneurship and Engineering Education: Dugald C. Jackson and the Cooperative Engineering Course, 1907–1932," *Technology and Culture* 29 (1988): 536–569; Larry Owens, "Vannevar Bush and the Differential Analyzer: The Text and Context of an Early Computer," *Technology and Culture* 27 (1986): 63–95; Larry Owens, "MIT and the Federal 'Angel': Academic R&D and Federal-Private Cooperation before World War II," *Isis* 81 (1990): 189–213; Alex Pang, "Edward Bowles and Radio Engineering at MIT, 1920–1940," *Historical Studies in the Physical and Biological Sciences* 20 (1990): 313–337; Karl Wildes and Nilo Lindgren, *A Century of Electrical Engineering and Computer Science at MIT, 1882–1982* (Cambridge, MA: MIT Press, 1985). For a discussion of the California Institute of Technology in the 1920s and 1930s, see Judith Goodstein, *Millikan's School: A History of the California Institute of Technology* (New York: W. W. Norton, 1991).

2. Robert H. Richards, *Ore Dressing* (New York: Engineering and Mining Journal, 1903); Heinrich Hoffman, "Notes on the Metallurgy of Iron and Steel" (unpublished manuscript, Stanford University Libraries, ca. 1900); Frank Hall Thorp, *Outlines of Industrial Chemistry: A Textbook for Students* (New York: Macmillan, 1898); Gaetano Lanza, "The Educational Process of Training an Engineer," *Technology Quarterly* 6 (1893): 173–180.

3. Francis Amasa Walker to "Mr. Lyon," March 24, 1894, collection 298, box 2, folder correspondence 1894, MIT Institute Archives and Special Collections, Cambridge, Massachusetts; Louis Derr to Henry Pritchett, June 8, 1903, and Samuel Woodbridge to Henry Pritchett, June 20, 1903, collection 85–44, MIT Institute Archives and Special Collections, Cambridge, Massachusetts.

4. John R. Freeman, "Silas Whitcomb Holman," *Technology Review* 3, no. 1 (1901): 28–29; Arthur Noyes, "Report of the President," in *MIT Report to the President* (1908); Arthur Noyes, "Talk to First-Year Students," *Technology Review* 9 (1907): 5. Most MIT *Reports to the President* are now available online; they will be cited here by year as *MIT Annual Report*. See <http://libraries.mit.edu/archives/mithistory/presidents-reports.html>.

5. John W. Servos, *Physical Chemistry from Ostwald to Pauling: The Making of a Science in America* (Princeton, NJ: Princeton University Press, 1990).

6. Noyes, *MIT Annual Report* (1908); Noyes, *MIT Annual Report* (1911).

7. William Walker, "Chemical Engineering and Industrial Progress," *Journal of Industrial and Engineering Chemistry* 3 (1911): 286–292.

8. Dugald C. Jackson, "Electrical Engineering and the Public," *Proceedings of the American Institute of Electrical Engineers* 30 (1902): 1138; Dugald C. Jackson, "The Typical College Courses Leading with the Professional and Theoretical Phases of Electrical Engineering," *Science* 18, no. 466 (December 4, 1903): 710–716, on 711–712; William Walker, "What Constitutes a Chemical Engineer," *Chemical Engineer* 2 (1905): 1–3.

9. Arthur D. Little, "A Laboratory for Public Service," *Technology Review* 11 (1909): 19.

10. William Walker, "A Laboratory Course in Industrial Chemistry," *Technology Review* 6 (1904): 163–174; "Chemistry and Chemical Engineering," *Technology Review* 7 (1905): 293–295; Little, "A Laboratory for Public Service," 16–18, 22–24; William Walker, "RLAC," in *MIT Annual Report* (1909, 1911); William Walker, "The University and Industry," *Journal of Industry and Engineering Chemistry* 8 (1916): 63–65; Warren K. Lewis, "History of the RLAC," ca. 1921, AC13, box 12, folder 355, MIT Institute Archives and Special Collections, Cambridge, Massachusetts.

11. Henry Talbot to Henry Pritchett, September 3, 1903, collection 85–44, MIT Institute Archives and Special Collections, Cambridge, Massachusetts; Henry Talbot, "The Engineering Graduate: His Strength and His Weakness," in Richard Maclaurin, ed., *Technology and Industrial Efficiency* (New York: McGraw-Hill, 1911), 114–123; Henry Talbot, "Chemistry and Chemical Engineering," in *MIT Annual Report* (1901, 1905, 1908).

12. Henry Pearson, *Richard Cockburn Maclaurin* (New York: Macmillan, 1937); Herbert Baldwin, "An Interview with George Eastman," *Technology Review* 22

(1920): 74–77; "Eastman's Gifts," July 1, 1929, AC13, box 14, folder 412, MIT Institute Archives and Special Collections, Cambridge, Massachusetts.

13. Carlson, "Academic Entrepreneurship and Engineering Education"; "Report of the Visiting Committee of the Department of Chemistry and Chemical Engineering," December 6, 1915, AC13, box 5, folder 125, MIT Institute Archives and Special Collections, Cambridge, Massachusetts; Arthur D. Little to William Walker, May 21, 1916, AC13, box 21, folder 615, MIT Institute Archives and Special Collections, Cambridge, Massachusetts; Richard C. Maclaurin, *MIT Annual Report* (1916); William Walker, "A Master's Course in Chemical Engineering," *Technology Review* 18 (1916): 837–845.

14. Richard C. Maclaurin to Charles Stone, August 15, 1919, AC13, box 19, folder 550, MIT Institute Archives and Special Collections, Cambridge, Massachusetts; "The Technology Plan," *Technology Review* 22 (1920): 532–561; Richard Freeland, "The Technology Plan at MIT, 1920–1940: A Case Study of University-Industry Cooperation," in *Massachusetts Higher Education in the Eighties: Research and the Economy* (Boston: University of Massachusetts Press, 1986), 8–23; Pearson, *Richard Cockburn Maclaurin*.

15. Charles Locke to Richard C. Maclaurin, October 21, 1919, collection AC13, box 13, folder 379, MIT Institute Archives and Special Collections, Cambridge, Massachusetts; Richard C. Maclaurin to Harry Goodwin, October 31, 1919, collection AC13, box 9, folder 265, MIT Institute Archives and Special Collections, Cambridge, Massachusetts; Irenée du Pont to Haskell, November 22, 1919, collection AC13, box 9, folder 246, MIT Institute Archives and Special Collections, Cambridge, Massachusetts; Everett Morss, "Technology Plan Contract," January 17, 1924, collection AC13, box 14, folder 411, MIT Institute Archives and Special Collections, Cambridge, Massachusetts.

16. "The Technology Plan," 532–561; Charles Norton, "Massachusetts Institute of Technology Plan of Industrial Cooperation," *Journal of the American Ceramic Society* 7 (1924): 248.

17. Arthur Noyes, quoted in Freeland, "The Technology Plan at MIT."

18. Arthur Noyes to Richard C. Maclaurin, April 23 and November 17, 1919, collection AC13, box 15, folder 437, MIT Institute Archives and Special Collections, Cambridge, Massachusetts.

19. William Walker, "The Technology Plan," *Science* 51 (April 9, 1920): 357–359, on 359.

20. William Walker to Administrative Committee, November 17, 1920, collection AC13, box 21, folder 616, MIT Institute Archives and Special Collections, Cambridge, Massachusetts; "The Reunion: The Present and Future of Technology," *Technology Review* 22 (1920): 383–384.

21. William Walker to Administrative Committee, July 22, 1920, and Charles Norton to Samuel Stratton, February 13, 1929, collection AC13, box 25, folder 176L, MIT Institute Archives and Special Collections, Cambridge, Massachusetts; Henry Talbot, *MIT Annual Report* (1921); Charles Norton, "Five Years of the Technology Plan," *Technology Review* 26 (1924): 78–81; Charles Norton, "Division of Industrial Cooperation and Research," in *MIT Annual Report* (1924); "Conference between Heads of Departments and the Corporation on the Division of Industrial Cooperation and Research," February 23, 1927, collection AC13, box 21, folder 615, MIT Institute Archives and Special Collections, Cambridge, Massachusetts.

22. Leroy Foster, "Sponsored Research at MIT, 1900–1968," vol. 1 (unpublished manuscript, 1984), MIT Institute Archives and Special Collections, Cambridge, Massachusetts; Charles Norton to Samuel Stratton, February 13, 1929; Vannevar Bush, oral history, collection MC143, 29, 154, 529–573, 606–609, MIT Institute Archives and Special Collections, Cambridge, Massachusetts; Maurice Holland, *Industrial Explorers* (New York: Harper and Brothers, 1928), 76–91.

23. Charles Norton, "DICR," in *MIT Annual Report* (1924); William Walker, "Research Laboratory of Applied Chemistry," in *MIT Annual Report* (1919); Warren K. Lewis to Samuel Stratton, April 21, 1925, collection AC13, box 12, folder 356, MIT Institute Archives and Special Collections, Cambridge, Massachusetts.

24. "Report of the Visiting Committee of the Corporation," 1923, collection AC13, box 5, folder 128, MIT Institute Archives and Special Collections, Cambridge, Massachusetts; Dugald Jackson to Everett Morss, December 22, 1924, collection AC13, box 14, folder 411, MIT Institute Archives and Special Collections, Cambridge, Massachusetts; "Visiting Committee Reports: 1. Electrical Engineering and Physics," 1925, collection AC 13, box 5, folder 129, MIT Institute Archives and Special Collections, Cambridge, Massachusetts; Dugald C. Jackson to Gerard Swope, December 22, 1926, collection AC13, box 19, folder 564, MIT Institute Archives and Special Collections, Cambridge, Massachusetts; Gerard Swope to Samuel Stratton, January 2, 1926, collection AC13, box 19, folder 564, MIT Institute Archives and Special Collections, Cambridge, Massachusetts.

25. George Harrison, "Karl Compton" (unpublished manuscript, 1961), collection MC105, MIT Institute Archives and Special Collections, Cambridge, Massachusetts; Karl Compton to Gerard Swope, collection AC4, box 236, folder G, MIT Institute Archives and Special Collections, Cambridge, Massachusetts.

26. "Report of Massachusetts Institute of Technology Visiting Committee of the Corporation on the Division of Industrial Cooperation and Research," March 9, 1927, AC13, box 16, folder 138, MIT Institute Archives and Special Collections, Cambridge, Massachusetts; Charles Norton to Samuel Stratton, April 26, 1927, December 18, 1928, and February 13, 1929, AC13, box 25, folder 176, MIT Institute Archives and Special Collections, Cambridge, Massachusetts; Frederick Keyes, "Memorandum regarding Outside Work, Chemistry Department," January 1931, AC4, box 217, folder 2, MIT Institute Archives and Special Collections, Cambridge, Massachusetts; Karl Compton, *MIT Annual Report* (1931).

27. Karl Compton, *MIT Annual Reports* (first half of the 1930s).

28. Harrison, "Karl Compton"; Philip M. Morse, *In at the Beginnings: A Physicist's Life* (Cambridge, MA: MIT Press, 1977); *MIT Annual Reports* (first half of the 1930s).

29. *MIT Annual Reports* (first half of the 1930s).

30. "Report of the Visiting Committee of the Corporation," *Technology Review* 35 (1932): 24–26; Karl Compton, "To Junior Members of the Staff," November 2, 1934, collection AC64, MIT Institute Archives and Special Collections, Cambridge, Massachusetts; Servos, "The Industrial Relations of Science."

31. Foster, "Sponsored Research at MIT"; Larry Owens, "MIT and the Federal 'Angel'"; Robert Kargon and Elizabeth Hodes, "Karl Compton, Isaiah Bowman, and the Politics of Science in the Great Depression," *Isis* 76 (1985): 301–318.

32. Karl Compton, "Memorandum of Relations of MIT to GE Company," June 30, 1950, AC4, box 85, folder 2, MIT Institute Archives and Special Collections, Cambridge, Massachusetts; Minutes of the Committee on Patents, November 28, 1938, collection AC64, MIT Institute Archives and Special Collections, Cambridge, Massachusetts.

DEBORAH DOUGLAS

In February 1941, war was raging in Europe, North Africa, and the Far East. Although the United States was still officially neutral, the Roosevelt administration was directing as much military material and equipment as was possible politically to friendly nations under siege. Most MIT scientists and engineers—like their colleagues across the country—believed that the United States would be going to war sooner rather than later, and they were already mobilizing to contribute to the nation's upcoming effort. But none of them could have anticipated how this war would reshape not only the physical campus and research agendas but also the way that MIT educated its students.[1]

That same month, nineteen-year-old Herbert Goldstein took steps to achieve a personal dream: he applied to do graduate work in physics at MIT. The dark-haired, bespectacled, and brilliant young man had already graduated Phi Beta Kappa from the City College of New York the previous year, and had won a teaching fellowship in the college's Department of Physics.[2] But MIT was his goal, and just ten days after the Institute received his application, Goldstein opened a return letter from Professor John Clarke Slater, chair of MIT's Department of Physics. Slater wanted Goldstein to know that he was on the "preferred list of candidates" in case Goldstein should receive offers from other graduate schools.[3] On April 1, Goldstein heard from MIT's director of admissions: yes, he had been admitted to MIT, but his application for a tuition scholarship was denied.[4] Goldstein immediately sent a telegram to Slater: "Wish to know if rejection of scholarship application sent by Prof. Thrasher is an error." Slater replied: "Regret funds insufficient for tuition scholarship in spite of high recommendation by department."[5]

Considering this painful disappointment, one can imagine young Goldstein's euphoria when he received different news on April 8: another
applicant had relinquished the MIT department's offer, and so if Goldstein
wanted it, a tuition scholarship and a teaching fellowship with a salary of
five hundred dollars for the 1941–1942 academic year was his.[6] Goldstein
wired the next day: "Definitely accept offer stated in your letter of April
8. Thank you."[7] Yet in the heady days that followed, Goldstein discovered
that he had an excruciating decision to make. As an Orthodox Jew, he
observed the Sabbath. Suppose his teaching schedule included Saturday
classes. Could he request a schedule change? Would merely acknowledging
his religion be problematic? This had not been a problem at City College
with its large number of Jewish students and faculty members. Nor would
it have been a problem at the University of Rochester, where Goldstein
had also been admitted; the prominence of Rochester professor Victor
Weisskopf, a deeply devout and observant Orthodox Jew with whom
Goldstein had been corresponding, mitigated the chance of difficulties. But
in conversations with friends and colleagues, the issue of discrimination
against Jewish students and faculty, which was so widespread in many elite
U.S. universities during this time, quickly came to the fore.[8]

Herbert's father, Harry, wrote to their rabbi, Israel Upbin, for advice
and assistance.

Let us suppose he writes a request now asking Professor Slater, Chairman of the
Physics Department, (who, incidentally, was directly responsible for the appointment) to program Herbert's work so that the Saturday will be free for religious
purposes. Understandably, upon receiving such a request, Professor Slater may be
seized with misgivings. Perhaps this appointee is of a distorted pattern, emotionally
unstable, of neurotic leanings, anti-social, a non-conformist, a moody fellow,
unpleasant to have around—in a word, a fanatic, who will no doubt give annoy-
ance and cause inconvenience.[9]

Harry Goldstein continued that his son felt that waiting until the fall to
make his request was utterly dishonorable. Their hope was that the rabbi
might write to Harvard professor Nathan Isaacs, prominent among Orthodox Jews for his public adherence to Jewish law despite the well-known
and deep-rooted anti-Semitic sentiments of Harvard President Abbott Lawrence Lowell. Perhaps the rabbi could persuade Isaacs to vouch for Herbert
in a special appeal to MIT's John Slater. Rabbi Upbin sent a letter to Isaacs,

FIGURE 4.1

At age nineteen, Herbert Goldstein was ecstatic when MIT admitted him in 1941 for graduate work in physics. All new students to MIT were required to submit portraits. The annotation ("Compton's successor!!") reveals something of Goldstein's ambitions.

Source: Courtesy of the MIT Museum.

but the Harvard professor was extremely ill (he died in December 1941) and did not respond to the request.[10]

Herbert Goldstein's determination to become a scientist and contribute to the war effort in his chosen field of physics overcame his fear of a hostile reaction, and so on May 5, he responded to Slater's inquiry about his preferences for a teaching assignment. "My dear Professor Slater," Goldstein wrote, "I desire to say I shall be happy to serve at any post, though there is the thought that work in the advanced laboratories may offer richer experiences." At this point, one can almost feel Goldstein drawing in his breath. "The request I do wish to make is to have your offices program my assignment so as to leave Saturdays free. This I ask in order that I may observe the Sabbath. I will gladly take on additional duties to make possible such an arrangement."[11] Slater responded immediately. "Dear Mr. Goldstein: Thank you for your letter of May 5th. I will be very glad to arrange your schedule so that you will have no duties on Saturday."[12] When the teaching assignments were made the following month, Goldstein saw that he would be working only the 250 hours per term expected of all teaching fellows in the Department of Physics. What he could not predict on that day in May was that by choosing MIT, he would be matriculating into a university already in high gear and more than ready to help him fulfill his personal aspiration to help the war effort.[13]

"TECHNOLOGY" GOES TO WAR

On September 19, 1940, MIT President Karl Taylor Compton attended a top-secret meeting in Washington, DC, with scientists from Britain and the United States at the forefront of radar research. Compton already belonged to the newly created National Defense Research Committee (NDRC) and headed its Division D, which dealt with instruments and controls, including radar. Alfred Loomis—a Wall Street financier, scientific visionary, life member of MIT's Corporation, and chair of the NDRC's microwave group—hosted the clandestine gathering under the guise of a party at the Wardman Park Hotel. Late that evening, the British guests opened a small wooden box and held up a compact black metal device that looked something like a hockey puck with tubes and wires attached. This device, Edward Bowen and John Cockcroft quietly announced, was a cavity magnetron that could generate ten-centimeter microwave pulses at ten kilowatts of power. It worked, they said, but with Britain at war with Germany, the scientists expressed the government's concern that their country did not have the necessary resources to undertake the full-scale research program needed to develop this promising new technology for the military's arsenal.[14]

This was the essence of the plea they were making to the U.S. scientists: England was prepared to give the United States one of the most important technologies ever discovered in exchange for full assistance in its development. For everyone present in the room, it was an unforgettable moment. All were thinking, if microwave radar could be developed, it would have amazing military applications. With waves of this length, the military could detect enemy aircraft, ships, or bombs at a far greater range with greater

FIGURE 4.2

Edward Bowen (in chair, left), Lee DuBridge (center), and Isador Rabi (right) with the ten-centimeter cavity magnetron. Bowen was a key figure in the British development of microwave radar and the person who carried the original magnetron to the top-secret meeting with U.S. officials in September 1941. The MIT Radiation Laboratory was born out of that meeting, with DuBridge as its director and Rabi as its lead researcher.

Source: Courtesy of the MIT Museum.

accuracy, and with less interference from ground reflection, clouds, or sea waves. Because the smaller the wave, the smaller the equipment, this new cavity magnetron device suggested that special radar sets could be developed and placed aboard airplanes and ships. This would be a tremendous military advantage—perhaps *the* most critical technological advantage—in the war.[15]

The United States was not yet at war, but there were few isolationists within the U.S. scientific-technological community. The rest of the country might be consumed with election debates about the risk of the United States being sucked into the European conflagration, but the four

Americans at the Wardman Park Hotel that evening did not express the slightest hesitation. They assured their British colleagues that work would start immediately.[16] The Army and Navy would focus on long-wave radar systems (they had already begun work on these technologies, which were seen as having played a successful role in the Battle of Britain earlier that year), while the NDRC would concentrate on the promising though unproven microwave technology just disclosed. But where would the NDRC set up its new laboratory? For microwave radar technology to advance rapidly, there needed to be a central enterprise bringing together military personnel with civilian experts from industry and academia.

On October 17, 1940, Compton was back in Washington for a meeting with Vannevar Bush at the Carnegie Institution on the corner of 16th and P Streets. Until 1938, Bush had been at MIT, a professor and Vice President and Dean of Engineering. Now he was president of the Carnegie Institution and chair of the NDRC. Joining Compton and Bush that morning were financier Alfred Loomis, Edward Bowles (MIT Professor of Electrical Engineering), and Frank Jewett (President of the National Academy of Sciences and Chair of the board of Bell Laboratories). The purpose of the meeting was to decide on a location for the microwave lab. For various reasons, it had proven impossible to host such a lab in Washington (Bush and Loomis's first choice); although two West Coast universities had offered space, they were deemed too far from all the key partners; and although it remained unstated during this meeting, Bush and Loomis adamantly opposed Jewett's earlier proposal for Bell Laboratories to manage the effort. They had been incensed—perhaps naively so—by Jewett's assertion that MIT lacked the ability to manage such a complex undertaking. Jewett's proposal had been rejected the day before, and yet he gamely participated in the second meeting. During the following years, Bell Laboratories would never stint its resources, and without its involvement, it is not clear the new lab would have been as successful. Still, the decision not to locate the facility at Bell did somewhat sour Jewett's personal feelings toward MIT, his alma mater, for the rest of his life.[17]

The two questions that Bush, Loomis, Bowles, and Jewett put to Compton that morning were simple but urgent: Did MIT have ten thousand square feet of available lab space? Could MIT arrange for access to a facility at East Boston Airport (now Logan International Airport)? Compton telephoned his assistant, James Killian, with the first question.

After scrambling for a few hours, Killian called back to say that if Bowles's laboratory could be vacated, they would have a place to start. That must have produced some laughter in the group in Washington, but Bowles agreed. The following week, the NDRC approved the plan officially, and a contract for $455,000 (nearly $7 million in 2008 dollars) was signed to fund a year's operation and a staff of about fifty. That was a remarkable sum, almost 14 percent of the Institute's total budget of $3.3 million ($50 million in 2008 dollars) for the 1939–1940 year. But none of this was as remarkable as what the new microwave laboratory was about to become.[18]

Ernest O. Lawrence—an NDRC microwave committee member and a star physicist at the University of California—declined the directorship of the new lab; he wanted to continue work on his giant cyclotron, but he played an essential role in shaping the new lab. It was Lawrence who recruited another star—his former student, Lee DuBridge, Chair of the University of Rochester's Department of Physics and Dean of the Faculty of Arts and Sciences—to be head of the lab. DuBridge did not need much persuasion. He later recalled "that if Lawrence was interested in the program, that was what I wanted to be in."[19] He accepted the post on October 16, 1940. This was how Bush liked to work: identify the challenge, find the right person, get him (it was almost always a "him") to work immediately, and sort out the bureaucratic details later. It would be the model for how the new lab would recruit most of its staff and was based on Bush's experience working as an MIT administrator under Compton to build up key new departments at MIT during the 1930s.

Lawrence and DuBridge began to recruit the best minds in physics to join the lab. By November 6, Isador Rabi (Columbia) and Wheeler Loomis (Illinois) had joined the team. Twenty physicists in all were able to start by December 1. DuBridge wanted the keenest scientific minds to attack the theoretical questions of microwave radar. If together they could figure out why the British cavity magnetron worked, the development of the technology for military use, he reasoned, would proceed more quickly and to greater effect. Not surprisingly, this philosophy resonated with Compton. For Compton, the reason for placing the new lab at MIT was to be of service to the nation. But he probably also had some inkling that it might fulfill his ambitions for transforming MIT to a degree he had not initially conceived. The new Radiation Laboratory (so named to disguise the real research objective of the facility) was already shaping up to be an intense yet dynamic research environment.[20]

When Goldstein arrived on campus the following fall, he knew nothing about the Rad Lab (as everyone called it), but even as he was settling into his dormitory room on the sixth floor of MIT's Graduate House, workers were racing to finish construction on a thirty-eight-thousand-square-foot building (Building 24), right behind the famous Dome, to house the rapidly growing cadre of physicists, engineers, technicians, and other specialists. MIT had already signed fifty-five defense contracts worth $3.8 million (almost $58 million in 2008 dollars) and employed 466 staff members, including 320 scientific personnel. Seventy of the scientific staff members—including Goldstein's adviser, Slater—were MIT employees who now split their time between teaching and research duties at the Rad Lab. At the same time, a new laboratory for chemical engineering was being built with funds from the Army's Chemical Warfare Service. The two buildings eliminated quite a few parking spots (as controversial then as now) and required relocation of the outdoor running track.

The draft had not yet had an impact on the student body (the policy was to give deferments to men undergoing technical training), but many, especially in the admissions office, wondered when this might change. Students began to notice parts of the campus being shut off, and more and more workers appeared. While student enrollments remained about the same, MIT was beginning to host special training courses for the military and a range of government agencies in fields including meteorology, aeronautical engineering, and chemical engineering. There was a new Radar School. At the All-Tech Smoker to welcome incoming members of the class of 1945, President Compton noted that they would be asked to make some minor sacrifices—but for good cause. "No other issue or activity now compares in importance with the defeat of Hitler and the antidemocratic, anti-humanitarian aspects of his Nazi movement."[21]

When the Japanese attacked Pearl Harbor on December 7, 1941, Compton again spoke to the students, urging them to continue their studies. He noted that many students had abandoned their studies to sign up for military service during World War I. Their intent was noble, he said, but it did not help the nation as much as if they had finished their degrees, and contributed their scientific and technical expertise to the cause. Some things did change after Pearl Harbor—most significantly, a new accelerated schedule started that winter. The academic term was shortened from fifteen to eleven weeks for graduating seniors in the class of 1942. Although Compton and many faculty members were opposed in principle to rushing the educational experience, the Academic Council of MIT voted in June 1942 to institute a summer term that allowed undergraduates to graduate in just under three years instead of four. Students also noted the substitution of "war emergency subjects in place of many of the non-professional subjects which are normally desirable for cultural breadth and intellectual recreation." Soon blackout shades, air-raid drills, collections for War Bonds and the Red Cross, and even special recycling programs became common. The masthead for the student newspaper, *The Tech*, now included the mantra, "Let's Set the Rising Sun." The entire MIT community, in concert with the nation, turned to the single goal of winning the war.[22]

While a military atmosphere began to pervade all of campus, it was particularly significant for certain departments. Slater characterized the change as "the invasion of the Institute by physicists from all over."[23] But these were civilian scientists. By contrast, the Department of Aeronautical Engineering was teeming with military personnel. Day in and day out, MIT professors taught Army and Navy officers the fundamentals of aeronautical engineering, instrumentation, and meteorology. The newly dedicated MIT Wright Brothers Wind Tunnel started to run double shifts to accommodate the demand from the aircraft industry. Perhaps earlier than other MIT departments, aeronautical engineering felt keen pressure to increase the number of graduates. In a May 1940 speech to Congress, President Roosevelt had challenged the nation to increase its production capacity to at least fifty thousand airplanes a year. For an industry capable of producing only one-tenth that number, this created an extraordinary demand for aeronautical engineers. Universities like MIT would need to expand their teaching efforts even as more and more requests for research work poured in.[24]

BIRTH OF THE MILITARY-INDUSTRIAL-UNIVERSITY COMPLEX

Professor Charles Stark Draper, or "Doc" as everyone in the Department of Aeronautical Engineering called him, was the king of instruments. Behind locked doors on the second floor of the Guggenheim Aeronautical Laboratory (also known as Building 33), he and a clutch of students who

FIGURE 4.3
Life at MIT changed within days of the Japanese attack on Pearl Harbor in December 1941, as the headlines from the student newspaper, *The Tech*, reveal.
Source: Courtesy of *The Tech*.

comprised the MIT Confidential Instruments Development Laboratory were completely focused on solving a critical problem for the U.S. Navy: fire control. With increasingly fast aircraft, the challenge of hitting an intended target was more difficult than ever. In June 1940, Doc entered into a financial partnership with Sperry Gyroscope in order to transform his uniquely designed aircraft turn-indicator into a lead computing gunsight. He desperately needed funds to continue the work, and Sperry wanted to keep tabs on university research.[25]

In December 1940, the first prototype—"Doc's Shoebox"—was working, and demonstrations to various visiting military officials and scientists produced considerable interest. Further development produced what is now viewed as a revolutionary new type of gunsight, but initially Sperry found it difficult to manufacture the devices. On the surface, the problem seemed to be that Doc based his invention on technologies that threatened to make Sperry's existing equipment obsolete, but the real problem was likely one of technology transfer. Doc and his students were forever tinkering, tweaking, and recalibrating. As the Navy placed ever larger orders (twelve, fifty, and then twenty-five hundred units) for the instrument, Sperry engineers and executives despaired of Doc's attitude that only his students were qualified to work on this project. Essentially Doc wanted Sperry to reproduce the identical environment and culture of his MIT lab. The problem became unbearably acute in February 1942 when the Navy increased its order to twenty-one thousand units, with a minimum delivery of a thousand per month.[26]

Draper and Sperry eventually achieved a workable arrangement, in part through their partnership with feedback-control pioneer Professor Gordon Brown of the MIT Servomechanism Laboratory. The Navy bought a total of eighty-five thousand Mark 14 gunsights—"fire control for the masses" because the device was easy to use and sufficiently accurate for its wartime application.[27] Draper and MIT would ultimately win considerable recognition for this and many other contributions to the war effort, but the significance of the work extended beyond the gunsight itself to the new kind of partnership forged. The Navy (and Army) had not been entirely happy with civilian scientists in control of military technology research, as was the case with the NDRC. For example, the NDRC had intellectual oversight of the Rad Lab; MIT was legally just the lab's host. The partnership between Draper's lab, Sperry, and the Navy was different: the lab was entirely separate from the NDRC's Fire Control Committee and was governed by an entirely different set of contractual relations. Further, MIT quickly recognized that Draper's lab represented a new pedagogical enterprise.

All across MIT, this new hybrid model of a university laboratory took hold. These labs blended instruction with real-world problem solving (although the balance varied with the faculty member in charge). MIT President Compton became enthusiastic about the ways that "specific research projects undertaken for government and industry by the group have led to advances in the art which will be reflected in the educational program."[28] Particularly talented students were immediately tapped to teach as well as work on or even lead key projects. For example, Doc Draper turned to his graduate student Robert Seamans (who would later become deputy administrator of NASA and secretary of the U.S. Air Force) to take on a problem that Sperry had asked Draper to solve. Although this project originated in industry, Draper informed Seamans that this work on vibration-measuring equipment would serve as his master's thesis.[29] Professor Brown was so impressed with the intellect and managerial skill of his student Jay Forrester that he quickly made Forrester an assistant director of the Servos Lab. Forrester's early successes resulted in Brown's assigning him to a special Navy project to build a new kind of flight simulator—an effort that Forrester would transform into Project Whirlwind to build the first real-time digital computer.[30] Slater, reflecting on a similar situation in the Rad Lab, put it this way: "That laboratory was an extraordinary assemblage of very able and intelligent people who had almost no experience with the problems they were working on. Many of them were very young, pulled out of graduate school. The fact that they could get so far was described later in the war as the 'miracle of children.'"[31]

Herbert Goldstein was one of those "children." Before the end of his first year at MIT, he had passed the written portion of his comprehensive exams, and Director DuBridge was eager to see him join the Rad Lab. By fall 1942, Goldstein was working part-time at the Rad Lab *and* starting work on his thesis. His adviser, Slater, began to push him toward problems that would contribute to the collective understanding of magnetron theory. In a letter to his father, Goldstein described one meeting in late December 1942: "At Slater's suggestion, I'm extending some of his formulae to other problems of interest. By all rights, it should be perfectly straightforward. But already I'm running into singularities & infinities. And I don't see my error. Darn it! Nothing is ever clear cut for me!"[32] In February Goldstein went to his adviser with a new proposal. Slater approved it and told Goldstein he could try for the June graduation list, adding, "I'll be gone six weeks, so I'll be back when the time comes for writing up."[33]

Most doctoral students recall the final weeks before completing the first draft of a dissertation with mixed emotions. It is a challenging time intellectually, physically (Goldstein contracted German measles, which required a week's quarantine), and emotionally. Goldstein struggled with the calculations, but his letters reveal a gradual understanding and, more important, the support of his Rad Lab colleagues. "I'm still convinced the edge effects are the source of error," he wrote to his father, "and [physicist Julian] Schwinger agrees, but how to take them into account is another story."[34] Goldstein managed to finish the draft just under the wire; then there were the grilling of his readers, corrections, and a defense. He faithfully recounted it all in his letters, from near tears to handshakes. While he was happy about graduating, he struggled some with his future course and yearned for some career advice from Slater. But Slater's six-week trip to England turned into a three-month stay; he didn't return until well after graduation, and by then Goldstein, just twenty-one years old and still a bit dazed by the accelerated degree program he had just completed, had landed a full-time job with the Rad Lab's Theoretical Group.[35]

Such was the crazy nature of the wartime experience at MIT. Compton later reminisced that although the war period was unbelievably demanding, he missed the simplicity of the time. He wrote that "however difficult and numerous were the problems, there was just one criterion for their solution: 'Is this the most useful thing which MIT, or any one of us personally, can do to help win the war?'"[36] The students all grappled with the normal anxieties of wartime, especially those associated with the draft. Everyone was eager to serve, but most hoped they could use their brains to support the cause. Meanwhile, friends and family were being drafted and sent around the world, which sometimes made those on the home front feel guilty.

In fall 1944, Goldstein began writing a lengthy journal that he hoped to one day share with his father. Goldstein ached to respond to his father's questions about whether or not the Germans were pulling ahead of the Allies. For three years, he recorded descriptions of Rad Lab systems and other war technologies as well as deeply personal sentiments. "The specter of another war threatening to wipe out our entire civilization already looms large while yet we bury our dead and weep tears for those who suffered at Buchenwald." A future nuclear holocaust haunted Goldstein as much as the Nazi one experienced by the Jews. "There is some light, some hope. But only my faith in the Lord's way keeps me from being completely fatalistic. May it be that in the future years when I read this, I shall laugh at my fears."[37]

President Compton expressed similar concerns: "God grant that our future demonstrations of strength may be made with full effect under the

FIGURE 4.4

Group 43 (the Theory Group) worked on the tough theoretical problems for the entire Rad Lab. In addition to the group's pioneering work on noise and antennas, Julian Schwinger (at the blackboard) developed a waveguide theory that was considered especially important. Herbert Goldstein (second row, closest to camera) joined this group in 1943 after finishing his doctorate in only two years.
Source: Courtesy of the MIT Museum.

stress of a strong urge to be useful in peace, and never again under the dread compulsion of war." Yet Compton concluded with the firm assertion that "one safeguard against another war is to be strong in peace."[38]

The end of the war meant an end to an extensive array of activities across the campus: MIT had been involved in everything from gas masks and flamethrowers to freeze-dried foods and aerial nighttime photography. Twenty percent of the nation's physicists had worked at the Rad Lab; only the Manhattan Project, which developed the atomic bomb, employed more. Most went back to their home universities; some, like Goldstein, had to scramble to find new jobs. Most of the women who had worked as "computers" (before there were machines known as computers, the word referred to people—mainly women—whose job was to perform the thousands of calculations needed by engineers for their research analysis) or technicians at the Rad Lab went back to domestic life. The MIT course schedule reverted to the two-semester plan—and everyone at MIT, Compton insisted, should take a vacation as soon as it could be scheduled.[39]

Yet the war had permanently transformed the Institute in almost every way possible. It is interesting to note that one of MIT's most significant achievements did not involve any hardware. Because of Bush and Compton's prewar relationship, MIT became the pioneer in negotiating war contracts. Having signed some four hundred contracts, worth more nearly $100 million (roughly $1.2 billion in 2008 dollars), MIT was the single-largest wartime research and development contractor. Compton wrote dramatically, "The Institute spent on its war contracts as much money as it had spent on its normal activities during its previous 80 years of existence."[40] The extraordinary effect of these expenditures was what inspired Compton and others throughout MIT to look toward the postwar period through different spectacles.

On November 10–12, 1945, MIT hosted the Victory in Science exhibition, spread across the Great Court lawn. More than seventy-five thousand people attended the event, which gave the Boston-area public its first opportunity to see the wide variety of new technologies developed during the war. MIT's future course was inextricably tied to that of the military and industry, and this, in turn, would shape the nature of future scientific and technological discovery and practice. Just because the Rad Lab was shutting down did not mean that MIT wanted to discard its highly productive research environment or that the Department of Physics wanted to abandon its newly established preeminence in the field of microwaves. The Department of Electrical Engineering was thrilled that the Navy decided to continue funding Project Whirlwind. Doc Draper and in fact virtually the entire MIT faculty wanted to keep federal funds flowing into the Institute. From their perspective, MIT, while hardly the only American university to make vital wartime contributions, was among the most elite institutions to have demonstrated unique competence to solve "the problems of the Federal Government."[41] They had come to believe that, as Vannevar Bush put it, "science was an endless frontier."[42] The military-industrial-university complex was the new instrument of exploration.

This enthusiasm for science and technology elevated research over teaching, graduate students over undergraduates, new faculty over old, and even government over industry. These changes created tensions, as did the debates over more fundamental issues, such as whether or not the Institute should permit classified research on campus or the manner in which MIT would address the threat of communism. During the war, MIT had clearly embraced a nationalist ideology, but there were some who hoped the Institute might now take a more global perspective. Still, none of these debates altered certain facts: MIT was bigger; it had a better physical plant, a more talented faculty, a growing student population, and an enormous appetite for a permanent relationship with the federal government (especially in areas of national defense) to fund future research efforts.

These changes were not the result of a single decision made by Karl Compton one October afternoon in Washington, DC, in 1940; they were the product of collective decisions—personal as well as public—of some of the nation's most talented minds. MIT was shaped by a professor's willingness to grant a young man's request not to work on the Jewish Sabbath. It was shaped by the manner in which the war thrust highly motivated but inexperienced graduate students into the crucibles of invention and design, and then expected them to perform miracles. It was shaped by military officers who prized the certificates that implied they were alumni of MIT and who had formed new relationships directly with the faculty. It was shaped by professors who, fueled by astonishing stores of adrenaline, worked twenty-hour days, and shuttled around the world to advise and assist in any needed field (sometimes in their area of expertise, and sometimes in disciplines they were inventing along the way). It was shaped by administrators

FIGURE 4.5

MIT's "Victory in Science" exhibition in November 1945 was attended by seventy-five thousand people. Herbert Goldstein's annotated program reveals the extent of the displays, which gave the general public its first close look at the Institute's varied wartime contributions.

Source: Courtesy of the MIT Museum.

who rarely equivocated in the face of decisions of great consequence, and by the nation's willingness to place enormous bets on contributions that could be made by civilian scientists operating outside the normal constraints (and even conventional contracts) of government institutions. Each of their moments of decision culminated in a new type of laboratory, a new type of student, a new type of professor, and a new life of the mind. The MIT we know today—the MIT that is known around the world as a vital locus of technological creativity, scientific discovery, and education innovation—was born.

NOTES

1. Several accounts document the history of MIT during World War II. For an overview, see John Burchard, *Q.E.D.: M.I.T. in World War II* (Cambridge, MA: Technology Press, 1948).

2. In 2008, Goldstein's family donated an extraordinary collection of artifacts and papers to the MIT Museum. The collection includes almost daily correspondence between Herbert and his father, Harry, and is the most detailed description available of graduate student life at MIT during World War II. Although Goldstein's doctoral research concerned topics of vital interest to the war effort, his work was unclassified. Yet while describing everything from laundry to canned food that he wanted his father to send along, Goldstein wrote carefully about his investigations. His letters capture emotions, not specific equations. In September 1944, he started a journal that he hoped to share with his father some day. For three years, Goldstein wrote sporadically but in detail about every project he was connected with. The Goldstein family does not know if Herbert ever shared this journal with his father; they discovered it only recently among his papers. Goldstein enjoyed a distinguished career as a physicist after World War II. Author of one of the standard graduate textbooks, *Classical Mechanics* (Cambridge, MA: Addison-Wesley, 1950), he was a professor of nuclear science and engineering at Columbia University as well as a consultant for the Oak Ridge and Brookhaven National Laboratories. Goldstein was also a founding member and president of the Association of Orthodox Jewish Scientists.

The Herbert Goldstein Collection (2009.006) is located at the MIT Museum, Cambridge, Massachusetts. Documents from the collection are cited here (as HG) to indicate authorship, date, document number, and page number; most of the letters, however, are short and unpaginated.

3. Victor Weisskopf to Herbert Goldstein, March 31, 1941, HG003. Goldstein had transcripts sent to Yale, Rochester, Cornell, and Princeton in addition to MIT. No correspondence with Yale, Cornell, or Princeton is in this collection.

4. John K. Ackley to Herbert Goldstein, February 25, 1941, HG020; B. A. Thresher to Herbert Goldstein (acceptance letter), April 1, 1941, HG004; B. A. Thresher to Herbert Goldstein (fellowship application rejection), April 1, 1941, HG005.

5. Herbert Goldstein to John C. Slater (draft telegram), April 2, 1941, HG006; John C. Slater to Herbert Goldstein (telegram), April 3, 1941, HG007.

6. John C. Slater to Herbert Goldstein (telegram), April 8, 1941, HG009; John C. Slater to Herbert Goldstein, April 8, 1941, HG010.

7. Herbert Goldstein to John C. Slater (draft telegram), ca. April 9–10, 1941, HG011.

8. Herbert Goldstein to unknown (draft), ca. April 11–28, 1941, HG016. On discrimination in U.S. universities, see Marcia Graham Synnott, *The Half-Opened Door: Discrimination and Admissions at Harvard, Yale, and Princeton, 1900–1970* (Westport, CT: Greenwood Press, 1979); Victor Weisskopf to Herbert Goldstein, March 31, 1941, HG003.

9. Harry Goldstein to Israel Upbin, ca. April 11–28, 1941, HG015.

10. Israel Upbin to Nathan Issacs, April 28, 1941, HG019; Larry A. DiMatteo and Samuel Flaks, "Conservative Legal Realism: Nathan Isaacs, Jewish Law, and Modern Legal Theory," available at <http://works.bepress.com/larry_dimatteo/1>.

11. John C. Slater to Herbert Goldstein, April 11, 1941, HG017; Herbert Goldstein to John C. Slater, May 5, 1941, HG018.

12. John C. Slater to Herbert Goldstein, May 7, 1941, HG021.

13. John C. Slater to Herbert Goldstein, June 20, 1941, HG022.

14. There are several accounts of this meeting. The most reliable popular account is Robert Buderi, *The Invention That Changed the World: How a Small Group of Radar Pioneers Won the Second World War and Launched a Technological Revolution* (New York: Simon and Schuster, 1996). Jennet Conant's *Tuxedo Park: A Wall Street Tycoon and the Secret Palace of Science That Changed the Course of World War II* (New York: Simon and Schuster, 2002) tells this story from the perspective of a key participant.

15. James Phinney Baxter III, *Scientists against Time* (Boston: Atlantic Monthly Press, 1946), 143–146.

16. Daniel J. Kevles, *The Physicists: The History of a Scientific Community in Modern America* (New York: Vintage Books, 1978), 287–289.

17. Buderi, *The Invention,* 45–46; Burchard, *Q.E.D.,* 219–220.

18. James R. Killian Jr., *The Education of a College President: A Memoir* (Cambridge, MA: MIT Press, 1985), 22–24; Burchard, *Q.E.D.,* 220.

19. Lee DuBridge, quoted in Henry E. Guerlac, *Radar in World War II* (1946; repr., New York: American Institute of Physics, 1987), 260.

20. Buderi, *The Invention,* 46–47. Both Lawrence's lab at Berkeley and the MIT facility intentionally used the name Radiation Laboratory to deceive Germany. In 1940, radiation work was considered to be the domain of the most theoretical of physicists and of no practical application to military needs. See also Kevles, *The Physicists,* 303.

21. "National Defense Spurs Institute Building Program," *The Tech* 61 (September 26, 1941); Burchard, *Q.E.D.,* 275–277; Karl Compton, "Report of the President," in *MIT Report to the President* (1941), 9; "Victory to Be Decided by Technical Superiority Dr. Compton Tells Frosh," *The Tech* 61 (September 30, 1941). Most *MIT Reports to the President* are now available online; they will be cited here by year as *MIT Annual Report.* See <http://libraries.mit.edu/archives/mithistory/presidents-reports.html>.

22. "Maintain Status Quo–K.T.," *The Tech* 61 (December 9, 1941); "Graduation Set for April 27; Compton Calls Open Meeting," *The Tech* 61 (December 17, 1941); "Text of Dr. Compton's Address," *The Tech* 61 (December 19, 1941); "The War-time Educational Program at the Massachusetts Institute of Technology," *Massachusetts Institute of Technology Bulletin* 77, no. 4 (1942): 30a–30e. The first appearance of "Let's Set the Rising Sun" was in the December 12, 1941, issue of *The Tech.* An editorial with the same heading appeared in the December 9, 1941, issue of *The Tech.*

23. John C. Slater, *Solid-State and Molecular Theory: A Scientific Biography* (New York: Wiley, 1975), 209.

24. *MIT Annual Report* (1942), 70–75; Shatswell Ober, "The Story of Aeronautics at M.I.T., 1895 to 1960," April 28, 1965, Department of Aeronautics and Astronautics, General History, MIT General Collection, MIT Museum, Cambridge, Massachusetts; Shatswell Ober, "The Wright Brothers Memorial Wind Tunnel (A Fragment of History of Aeronautical Engineering at M.I.T.)," Department of Aeronautics and Astronautics, General History, MIT General Collection, MIT Museum, Cambridge, Massachusetts.

25. Michael Aaron Dennis, "A Change of State: The Political Cultures of Technical Practice at the MIT Instrumentation Laboratory and the Johns Hopkins University Applied Physics Laboratory, 1930–1945" (PhD diss., Johns Hopkins University, 1990) 360–362, 381.

26. Ibid., 362–380.

27. David A. Mindell, *Between Human and Machine: Feedback, Control, and Computing before Cybernetics* (Baltimore: Johns Hopkins University Press, 2002), 221–223.

Mindell cites records that include the following: U.S. Navy, "Bureau of Ordnance: Summary of Progress," August 1, 1945, RG74 ordnance status reports, box 7, 32, U.S. National Archives, Washington, DC; *US Naval Administrative Histories of World War II*, vol. 73, *Research and Development, Maintenance* (Washington, DC: Department of the Navy, 1947), 160–168, available in the Navy Department Library Rare Book Room, Washington, DC.

28. *MIT Annual Report* (1945), 18.

29. Robert C. Seamans Jr., *Aiming at Targets: The Autobiography of Robert C. Seamans, Jr.*, NASA history series, NASA SP-4106 (Washington, DC: NASA History Office, Office of Policy and Plans, National Aeronautics and Space Administration, 1996), 30–31.

30. Kent C. Redmond and Thomas M. Smith, *Project Whirlwind: The History of a Pioneer Computer* (Bedford, MA: Digital Press, 1980), 14–15.

31. Slater, *Solid-State*, 212.

32. Herbert Goldstein to Harry Goldstein, December 23, 1942, HG025.

33. Herbert Goldstein to Harry Goldstein, March 1, 1943, HG029.

34. Herbert Goldstein to Harry Goldstein, April 9, 1943, HG030.

35. Herbert Goldstein to Harry Goldstein, May 13, 1943, HG031. Because of Slater's absence, the Department of Physics assigned two new readers to evaluate Goldstein's thesis. This is yet another example of the unusual nature of education during the war.

36. Karl Taylor Compton, quoted in Burchard, *Q.E.D.*, v.

37. Herbert Goldstein to Harry Goldstein, September 15 to December 2, 1944, HG032a:1; Herbert Goldstein to Harry Goldstein, January 7, 1945 to January 27, 1947, HG032b:83, 102.

38. Karl Taylor Compton, in *MIT Annual Report* (1945), 30–31.

39. Ibid., 31.

40. Ibid., 30.

41. "Victory in Science," *Technology Review* 48 (December 1945): 112; Slater, *Solid-State,* 217–225; Christophe Lécuyer, "The Making of a Science-Based Technological University: Karl Compton, James Killian, and the Reform of MIT, 1930–1957," *Historical Studies in the Physical and Biological Sciences* 23 (1992): 153–180, on 153.

42. Vannevar Bush, *Pieces of the Action* (New York: William Morrow and Company, 1970), 41; Kevles, *The Physicists,* 347.

had paid mostly for massive projects such as radar research at MIT's Radiation Laboratory and dropped quickly after hostilities ended. By the time the Lewis committee began its investigations, federal spending for sponsored research at MIT had fallen 75 percent from its wartime peak. Yet even that level struck committee members as being too high. Writing in 1949, the Lewis committee warned that federal funding had already caused every aspect of the Institute (bureaucracy, enrollments, and so on) to swell beyond "optimum size."[8]

Little could the committee members know that just a few months after they filed their report, the United States would enter the Korean War and MIT would once again lead the nation's universities in wartime mobilization. As MIT President James Killian explained in his annual report for 1952, the "Korean conflict produced an abrupt and compelling demand upon the Institute and its staff to make their special competence further available to aid the rearmament program." "The volume of our research conducted under contract with the government has rapidly risen," Killian continued, "to a total larger than we would wish if we were free of emergency demands." The Institute's operating budget leaped 36 percent during the first year of the new conflict and another 31 percent the following year—the fastest rates of growth since World War II.[9]

Unlike the situation after World War II, the Institute's scale of operations did not contract after the Korean conflict wound down to its stalemate; the campus saw no demobilization. Instead, defense spending at MIT continued to climb exponentially well into the late 1960s. Adjusted for inflation, the volume of sponsored research doubled every six years between 1948 and 1968.[10]

Throughout the 1950s and 1960s, sponsored research accounted for roughly 80 percent of MIT's operating budget. Of course, sponsored research had played an important role at MIT since the early days of President Richard Maclaurin's Technology Plan of 1919, which aimed to improve relations with local industries. The Institute's Division of Industrial Cooperation continued to manage outside contracts throughout World War II and the Korean War even as the source of sponsorship shifted dramatically from the private sector to government. During the 1950–1951 academic year, for example, the division processed more than $15 million in research contracts (nearly $135 million in 2008 dollars). More than 96 percent came from the federal government, virtually all from the

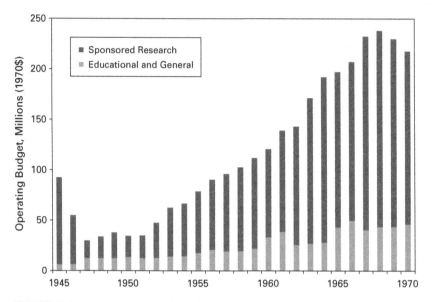

FIGURE 5.1

Annual operating budget for MIT, 1945–1970, in millions of inflation-adjusted, 1970 dollars. The bottom portion of each bar represents expenditures for educational and general operations; the top portion represents expenditures on sponsored research, dominated by defense-related federal agencies.

Department of Defense, the Atomic Energy Commission, and the National Advisory Committee on Aeronautics; just 3 percent came from industrial corporations. Twenty years later, the pattern looked remarkably similar. In 1969, 96 percent of sponsored research contracts came from federal agencies (again dominated by the Department of Defense, the Atomic Energy Commission, and NASA) and less than 1 percent from private industry. To reflect this trend, the Division of Industrial Cooperation was renamed the Division of Sponsored Research in 1956.[11]

Most of these funds went to special interdisciplinary laboratories and federal contract research centers—an arrangement initiated by two of MIT's renowned research facilities: the Instrumentation Laboratory and the Research Laboratory of Electronics.

Founded in 1934 by Charles Stark Draper, an aeronautical engineer, the Instrumentation Laboratory divided its time between military and commercial contracts before World War II. With the outbreak of fighting, Draper capitalized on his extensive experience with gyroscopes, developing a host of guidance and fire-control technologies for military use. Funded after the war almost entirely by military contracts, the Instrumentation Laboratory helped launch the space age. Working on inertial guidance systems for everything from ballistic missiles to the Apollo moon landings, Draper's lab quickly mushroomed into one of the largest and best-funded laboratories at MIT.[12]

The Research Laboratory of Electronics, established late in 1945, grew directly out of campus-based research on radar during World War II. MIT's Radiation Laboratory, known as the "Rad Lab," had been set up in 1940 to develop and produce radar for the military. Although the huge Allied effort on radar was headquartered there, the lab was not directed by MIT people. The Lewis committee joked in its 1949 report that the wartime influx of researchers from other institutions "conveyed the impression of an invasion from abroad," even if, gradually, "conciliatory changes characteristic of most extended foreign occupations began to occur."[13] Even before the war had ended, MIT faculty and administrators—spearheaded by physicists John Slater and Julius Stratton in conjunction with physicist and MIT President Karl Compton—aimed to give centers like the Rad Lab a more authentic MIT stamp. Military patrons, meanwhile, were eager to ease the transition from wartime operations without losing contact with the scientists and engineers who had proven so critical on projects like radar. Together they hammered out a deal in 1945, transferring the basic research wing of the wartime Rad Lab to the new Research Laboratory of Electronics, a joint venture between the Physics and Electrical Engineering departments. Major funding soon poured in from each branch of the military to support research on improved electronics, communications, radar, and missile telemetry. (And along with the first funds came tons of surplus Rad Lab equipment.)[14]

The Research Laboratory of Electronics (RLE) quickly became a template. The new Laboratory for Nuclear Science and Engineering, launched in December 1945 with inputs from five academic departments—physics, chemistry, electrical engineering, chemical engineering, and metallurgy— was explicitly modeled on the pattern of the electronics laboratory. As with the electronics case, military patrons stepped in to fund the new nuclear research. (The laboratory's founding director, physicist Jerrold Zacharias, tried to solicit funds from private industries as well, but few were interested.) MIT's first synchrotron—a powerful particle accelerator used to probe interactions among nuclear particles—came online in 1950 thanks to Navy support; MIT's experimental nuclear reactor, completed in 1958, was largely underwritten by the Atomic Energy Commission.[15]

The Pentagon played a proactive role in setting up other laboratories at MIT. Six months into the Korean War, U.S. Air Force officials approached MIT leaders, asking that the Institute further expand its already-expanded defense efforts. Out of these discussions came Lincoln Laboratory, an off-shoot of the Research Laboratory of Electronics, focused on air-defense systems. Even under the pressures of wartime, MIT administrators did not eagerly embrace the idea of taking on "Project Lincoln." One concern was that Lincoln would compete with the RLE. Such projects presented other "hazards" as well, MIT President Killian acknowledged in his annual report in 1952. "The conduct of large, classified military research projects imposes vexing and heavy burdens on any educational institution," Killian explained. "We look forward eagerly to the time" when projects like Lincoln "will be no longer necessary." All the same, he concluded, "an institute of technology has special resources which impose on it a responsibility in defense research different from many other kinds of educational institutions."[16]

Operations at Lincoln expanded quickly. Within a few years its staff had swelled to two thousand people, and in 1954 the laboratory, still under MIT auspices, moved to nearby Lexington, Massachusetts, close to the Hanscom Air Force Base. Following the detonation in 1949 of the first Soviet atomic bomb (and before the launch of Sputnik in 1957), most U.S. military strategists focused on the vulnerability of the United States to Soviet bombers. In response, Lincoln's first major project was the Semi-Automatic Ground Environment (SAGE), a network of radars and antiair-craft weapons connected via new digital computers, and distributed across the continental United States. At the center of the SAGE project was the famous Whirlwind high-speed digital computer, the brainchild of MIT researcher Jay Forrester.[17]

Dozens of smaller laboratories, scattered across MIT's campus, sprang up throughout the 1950s and 1960s on this model. Everything from the Gas Turbine Laboratory to the Naval Supersonic Wind Tunnel, the Aeroelastics and Structures Laboratory, the Servomechanisms Laboratory, the Center for Materials Science and Engineering, and more thrived on a steady diet of federal, mostly military, patronage. As Alvin Weinberg, physicist and director of the Oak Ridge National Laboratory, noted in 1962, it had

FIGURE 5.2
During the early to mid-1950s, scientists and engineers at MIT's new Lincoln Laboratory improved radar equipment and electronic computers, such as the Whirlwind computer (shown here), for a continental early-warning system to track incoming Soviet bombers.
Source: Courtesy of the MIT Museum.

become difficult "to tell whether the Massachusetts Institute of Technology is a university with many government research laboratories appended to it or a cluster of government research laboratories with a very good educational institution attached to it."[18]

HOW MANY STUDENTS? WHAT SHOULD THEY LEARN?

All that money paid for much more than laboratories and equipment. In fact, government patrons and campus administrators justified the skyrocketing expenditures in terms of neither gadgets delivered nor instruments installed, but rather numbers of students trained. In the wake of wartime projects such as radar and the atomic bomb, this kind of accounting became widespread. Politicians and pundits across the country adopted "scientific manpower" as their new rallying cry. Increasing the nation's "stockpile" of trained scientists and engineers—especially as the Cold War with the Soviets intensified—became an urgent priority. Policy began to reflect rhetoric, from the bankrolling of huge numbers of new federal fellowships to the rewriting of draft deferment policies, all in the name of getting more students into the nation's science classrooms and keeping them there. Reports (of dubious quality) that the Soviets were training twice as many scientists and engineers as the United States only fanned the flames. The manpower rush shaped educational policies across the nation's campuses. Once again, MIT led the way.[19]

Invoking the nation's "special needs," MIT's President Killian explained in his 1955 annual report how the Institute had been working to fulfill its own "special responsibilities." First and foremost, MIT had increased enrollments significantly. Even after the surge of World War II veterans had moved through MIT's classrooms (aided by the G. I. Bill), the Institute kept enrollments high—up 80 percent from the prewar levels. Undergraduate enrollments saw significant growth, but the fastest-rising share came from graduate students. In 1900, graduate students had amounted to 0.4 percent of MIT's total student population; by 1930, even after steady growth, they still accounted for less than 17 percent. Immediately after the war their numbers began to grow exponentially. By the late 1960s, fully half of all students at MIT were graduate students.[20]

The new interdisciplinary laboratories became efficient engines for pumping out graduates and were often justified in precisely those terms. When he was negotiating Navy support for the new Laboratory for Nuclear Science and Engineering late in 1945, for example, Admiral Harold Bowen of the Navy's Office of Research and Inventions stipulated to MIT President Compton that the laboratory should "integrate work under this contract with your educational program." "The more students who can use this opportunity for credit toward a degree," he continued, "the better will the Navy's ultimate interests be served in acquainting young men with our programs." Grants and fellowships from the other military services and the Atomic Energy Commission came with similar expectations.[21]

The plan worked. During its first decade, the Laboratory for Nuclear Science and Engineering served as the training ground for nearly three hundred graduate theses and one thousand undergraduate theses. Between 1946 and 1958, meanwhile, the Research Laboratory of Electronics supported six hundred student theses, and a comparable number were com-

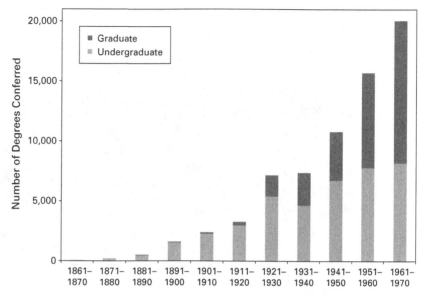

FIGURE 5.3

Number of degrees conferred by MIT per decade, 1861–1970. The bottom portion of each bar represents undergraduate degrees; the top portion represents graduate-level degrees, including master's, doctoral, and "advanced engineering" degrees.

pleted at the Instrumentation Laboratory. Bolstered by these giant new facilities, several of MIT's departments became the largest in the nation in their fields. During the 1958–1959 academic year, MIT awarded one in eight of all the doctoral degrees in engineering granted in the United States.[22]

The trend had been clear as early as the Lewis committee's deliberations in the late 1940s. The committee noted that the expansion of sponsored research on campus had "led to a preoccupation with graduate education at the sacrifice of attention to our undergraduate program." In fact, soon after the war, the campus hallways were filled with chatter about whether MIT should drop undergraduate education altogether. The suggestion proved so ubiquitous that the Lewis committee raised it no fewer than six times in its report—each time rejecting it soundly. Others had floated similarly radical ideas (likewise rejected by the Lewis committee), such as expanding the duration of undergraduate training to five or six years, to allow students sufficient time to absorb fast-expanding knowledge in

FIGURE 5.4
MIT's 350 MeV synchrotron, a particle accelerator operated by the Laboratory for Nuclear Science, probed interactions between elementary particles to study the nature of nuclear forces.
Source: Courtesy of the MIT Museum.

technical areas while still pursuing some form of well-rounded, general education.[23]

Even as it dismissed these brazen proposals, the Lewis committee could not duck the underlying question: What should all MIT students learn? Climbing enrollments forced the matter; any suggestion of returning to prewar routines faltered in the face of the Institute's lightning-fast expansion. The Lewis committee argued forcefully that MIT should never revert to the narrow, vocational-style technical training that had marked its earlier years. Rather, the Institute should bolster basic science to better foster creative thinking in its students, regardless of the specific topics or disciplines they would pursue. Striking that balance proved to be a perennial challenge. Other faculty committees revisited the topic in the 1960s and 1970s. Each time, they sought new ways to allow students greater flexibility in their choice of course work while also ensuring that undergraduates received appropriate depth within their major fields.[24]

A particular vision of the ideal student steered the Lewis committee's curricular recommendations. MIT needed to revamp its undergraduate curriculum, the committee urged, in order to train a new kind of engineer, one capable of meeting the special challenges of the postwar era. Radar and the atomic bomb had demonstrated beyond doubt that problems of science and technology were bound up inextricably with social and political concerns. In the nuclear age, the dividing line between "purely technical" and "politically relevant" projects blurred as never before. "The most important service M.I.T. can render to society," the committee concluded, would be to train its students to understand and appreciate the human consequences of their work. Such a lofty goal would redound to MIT's benefit as well: enhancing the humanities and social sciences would cure MIT graduates of their feeling of cultural "inferiority," and would appeal to a broader base of potential applicants who had been turned off by what looked to be MIT's one-sided technical training. So crucial did committee members consider this "humanizing" theme that they devoted 40 percent of their final report to it.[25]

The Lewis committee recommended meeting the postwar challenges by creating a new School of Humanities and Social Sciences at MIT, to share equal status with the Schools of Engineering, Science, and Architecture and Planning. Only by attracting top-flight faculty in the new areas, expanding the roster of required subjects, developing a healthy array of "civilizing" electives, and promoting cutting-edge graduate work in the humanities and social sciences could MIT successfully tackle the "most difficult and complicated problems confronting our generation." All MIT students, the committee urged, should have the opportunity to "develop an awareness of the interrelations of the scientific, technical, and literary cultures, and a sensitiveness to the diverse forces that motivate the thoughts and actions of people." In committee members' view, the proposed School of Humanities and Social Sciences should be no mere add-on. "We ask for more than a mechanical mixture of the conventional literary and technical cultures"; rather, they sought genuine "integration."[26]

MIT acted quickly to adopt this central recommendation of the Lewis committee's report. The School of Humanities and Social Studies was founded the following year, in 1950. (In 1959, its name was changed to the School of Humanities and Social Sciences, better reflecting the Lewis committee's original suggestion.) The new school began to grow almost immediately, thanks less to the idealism expressed in the Lewis report than to the same Cold War prerogatives that had thrown the rest of the Institute into overdrive. In 1950, researchers at MIT undertook a classified study known as "Project Troy," which focused on how to improve propaganda and psychological warfare techniques for the nation's Cold War arsenal. Project Troy was funded by the Office of Naval Research and the U.S. Department of State. Building on those relationships, MIT founded its Center for International Studies in 1951. The center, which continued to attract substantial funding from the Pentagon and the Central Intelligence Agency (CIA), was the first major addition to the brand-new School of Humanities and Social Sciences. Both Project Troy and the center certainly helped to integrate the School of Humanities and Social Sciences into the rest of MIT's operations—although probably not in the way that the Lewis committee had intended. Rather than help to dilute the military's influence on campus, the school settled quickly into the postwar pattern set by the Schools of Science and Engineering.[27]

A NEW ACCOUNTING OF VALUES AND MISSION

In 1970, more than twenty years after the Lewis committee delivered its recommendations, another blue-ribbon panel convened at MIT for another top-to-bottom inspection of where the Institute had been and where it should be going. The new committee, chaired by MIT mathematician Kenneth Hoffman, met amid a dramatically different setting: clashes between angry protesters and riot police had replaced the ticker-tape euphoria that had marked the end of World War II.

The Hoffman committee faced the painful collapse of the Cold War framework that had buoyed MIT since the late 1940s. Starting in the mid-1960s, Department of Defense officials began to reconsider whether open-ended basic research had produced the best return on their investment. The Mansfield Amendment to the Defense Appropriations Act of 1970 legislated what military planners had already been leaning toward on their own: it restricted military spending on the nation's campuses to projects of direct military relevance. There would be no more blanket grants to universities for the chief purpose of drumming up scientific recruits. Meanwhile, the fiscal demands of the escalating war in Vietnam, combined with the first glimmers of "stagflation"—rising inflation and stagnant economic growth—led to massive cuts in federal spending on education.[28] MIT, having become so dependent on federal grants, felt the shockwaves first. Budgets quickly began to slide. MIT faced a budget shortfall of $10 million between 1971 and 1973 (more than $50 million in 2008 dollars), its first deficit in decades. Enrollments likewise reached a plateau and then began to slip.[29]

The Hoffman committee's report in 1970, *Creative Renewal in a Time of Crisis*, returned to many of the Lewis committee's most important warnings and recommendations—warnings and recommendations that had largely been pushed to the sidelines during the Korean War and the long-drawn-out Cold War that followed. The Lewis committee had noted as early as 1949, for example, that the influx of sponsored research projects delivered significant dividends, most obviously in the form of world-class facilities and the unprecedented educational opportunities that those facilities could provide. Yet the committee had also warned of great costs associated with military patronage. Not only would the elephantine scale of operations necessitate huge expenditures of time and resources on accounting, bureaucracy, and similar "dissipations." The source of largesse proved as worrying as the overall scale of operations: too close an association in the public's mind between MIT and "war weapons" would not be healthy for the Institute.[30] Although the crisis was a long time in coming, it was finally pushed to the fore by the Vietnam War along with the outbursts of campus protests at MIT and across the country during 1969 and 1970. As the Hoffman committee noted in its report in 1970, "war-related research" at MIT, which had been "originally undertaken, well-nigh without dispute, in the service of the nation," suddenly took on a different cast in the light of "changing values."[31]

In the Hoffman committee's estimation, many other challenges facing MIT in 1970 bore an eerie resemblance to those highlighted by the Lewis committee decades earlier. Faculty and students alike were still frustrated by the curriculum. Too many subjects focused too narrowly on problem-solving techniques instead of fostering creativity and judgment. Students at both the undergraduate and graduate levels faced a dizzying array of specialized subjects without sufficient opportunity to integrate. The crush of students remained a major concern as well. During the 1969–1970 academic

FIGURE 5.5
MIT physics professor Anthony P. French teaching introductory physics to a packed crowd in the Compton Lecture Hall (room 26-100) during the 1964–1965 academic year.
Source: Courtesy of the MIT Museum.

year, more than half of all undergraduate credits were earned in classes with one hundred students or more. Large classes were especially problematic for first-year undergraduates: 93 percent of their credit hours in science subjects were spent in classes with one hundred students or more. Even by senior year, nearly 20 percent of students' credit hours were still spent in such bloated classes.[32]

Humanities and social sciences had not proven to be the panacea that the Lewis committee had hoped for either. The Hoffman committee reported with disappointment that rather than instill in students a robust appreciation for the social and political implications of their work, the school had become home to one more set of specialized academic disciplines confronting bewildered and overburdened students, cordoned off from other fields by narrow-gauge jargon and publish-or-perish demands on its faculty. Moreover, despite efforts by the Institute's administration to place humanities and social sciences on a par with science and engineering, those fields were still widely perceived—by students and faculty alike—as a lesser partner in the Institute's mission, to be shunted aside when the demands of "real" work beckoned. Like the Lewis committee before it, the Hoffman committee urged an immediate turnaround in the placement and prestige of humanities and social sciences within the MIT curriculum. Once again, the committee pointed to conditions beyond campus— the "social and political turmoil that swirls around us"—to argue for the need to foster "knowledge and values" on campus alongside technical know-how.[33]

Twenty years of runaway growth at MIT had produced dazzling innovations, contributed significant improvements to the nation's defenses, and introduced nearly thirty-six thousand graduates into the technoscientific workforce. But they had clearly taken their toll. "MIT has the appearance of great busyness and terrific efficiency, but not of great reflectiveness," the Hoffman committee concluded in its 1970 report. "The very fact that we have 'awakened' to find ourselves uncertain of our purposes and direction lends weight to this judgment." The sudden cessation of relentless growth offered the Institute a vital opportunity to pause and reassess its values as well as mission. "We have no intention of proposing that MIT adopt as an ideal an image of lolling ease and self-indulgent disorder," the Hoffman committee made clear. "But one must breathe."[34]

NOTES

1. Warren K. Lewis et al., *Report of the Committee on Educational Survey* (Cambridge, MA: MIT Press, 1949), 49.

2. Ibid., 4.

3. Ibid., 115.

4. Ibid., 14, 4.

5. Ibid., 16, 18. Scientists and administrators across the United States aired similar concerns throughout the late 1940s and 1950s. See David Kaiser, "The Postwar Suburbanization of American Physics," *American Quarterly* 56 (December 2004): 851–888.

6. Compare Dorothy Nelkin, *The University and Military Research: Moral Politics at M.I.T.* (Ithaca, NY: Cornell University Press, 1972), chap. 2; Roger Geiger, *Research and Relevant Knowledge: American Research Universities since World War II* (New York: Oxford University Press, 1993); Stuart W. Leslie, *The Cold War and American Science: The Military-Industrial-Academic Complex at MIT and Stanford* (New York: Columbia University Press, 1993); Rebecca Lowen, *Creating the Cold War University: The Transformation of Stanford* (Berkeley: University of California Press, 1997); David Kaiser, "Cold War Requisitions, Scientific Manpower, and the Production of American Physicists after World War II," *Historical Studies in the Physical and Biological Sciences* 33 (Fall 2002): 131–159.

7. Leslie, *The Cold War*, 235.

8. Lewis et al., *Report*, 15. On funding trends, see the annual *President's Report* for the years 1945–1949. Such reports (hereafter cited by year as *MIT Annual Reports*) are published each year as a special issue of the *Massachusetts Institute of Technology Bulletin* (Cambridge, MA: MIT Press). Copies are available in T171.M4195, MIT Institute Archives and Special Collections, Cambridge, Massachusetts, and online at <http://libraries.mit.edu/archives/mithistory/presidents-reports.html> (accessed December 20, 2008).

9. *MIT Annual Report* (1952), 29. Changes in operating budgets were calculated from data in the *MIT Annual Report* for the years 1950–1953.

10. Data in figure 5.1 compiled from the *MIT Annual Report* for the years 1945–1970. Adjustments for inflation were calculated using the consumer price index, maintained by the Bureau of Labor Statistics of the U.S. Department of Labor. See <http://data.bls.gov> (accessed June 1, 2008).

11. *MIT Annual Report* (1951), 234, 278–279, 312; *MIT Annual Report* (1952), 310; Wayne Stuart, *Facts about MIT* (Cambridge, MA: MIT Press, 1971), 67. On the Division of Industrial Cooperation and Maclaurin's vision, see Lewis et al., *Report*, chap. 4; Christophe Lécuyer, "MIT, Progressive Reform, and 'Industrial Service,' 1890–1920," *Historical Studies in the Physical and Biological Sciences* 26 (1995): 35–88; Lécuyer's chapter in this volume. On the name change to Division of Sponsored Research, see *MIT Annual Report* (1956), 27.

12. Leslie, *The Cold War*, 76–82; David Mindell, *Between Human and Machine: Feedback, Control, and Computing before Cybernetics* (Baltimore: Johns Hopkins University Press, 2002); David Mindell, *Digital Apollo: Human and Machine in Spaceflight* (Cambridge, MA: MIT Press, 2008).

13. Lewis et al., *Report*, 14.

14. *MIT Annual Report* (1946), 134–135; Leslie, *The Cold War*, 22–26.

15. Silvan S. Schweber, "Big Science in Context: Cornell and MIT," in *Big Science: The Growth of Large-Scale Research*, ed. Peter Galison and Bruce Hevly (Stanford, CA: Stanford University Press, 1992), 149–183; Leslie, *The Cold War*, 143–146, 156–157.

16. James Killian, in *MIT Annual Report* (1952), 31; Leslie, *The Cold War*, 32–33.

17. Leslie, *The Cold War*, 34–41. On SAGE and Whirlwind, see also Paul Edwards, *The Closed World: Computers and the Politics of Discourse in Cold War America* (Cambridge, MA: MIT Press, 1996), chap. 3; Atsushi Akera, *Calculating a Natural World: Scientists, Engineers, and Computers during the Rise of U.S. Cold War Research* (Cambridge, MA: MIT Press, 2008), chap. 5.

18. Alvin Weinberg, "The Federal Laboratories and Science Education," *Science* 136 (April 16, 1962): 30, as quoted in Nelkin, *The University*, 24. Also quoted in Leslie, *The Cold War*, 15.

19. Kaiser, "Cold War Requisitions"; David Kaiser, "The Physics of Spin: Sputnik Politics and American Physicists in the 1950s," *Social Research* 73 (Winter 2006): 1225–1252.

20. *MIT Annual Report* (1955), 12; Kenneth Hoffman et al., *Creative Renewal in a Time of Crisis: Report of the Commission on MIT Education* (Cambridge, MA: MIT Press, 1970), 29. Data in figure 5.3 are compiled from the *MIT Annual Report* for the years 1931–1970.

21. Harold G. Bowen to Karl Compton, November 8, 1945, quoted in Leslie, *The Cold War*, 144; see also Kaiser, "Cold War Requisitions."

22. Leslie, *The Cold War*, 97, 133–134; Geiger, *Research and Relevant Knowledge*, 68. On MIT's dominance in engineering doctoral training, see Stuart, *Facts about MIT*, 151.

23. Lewis et al., *Report*, 18. On the question of abandoning undergraduate education, see ibid., 15, 18–20, 25, 131. On expanding the duration of undergraduate training, see ibid., 20–21.

24. MIT Committee on Curriculum Content Planning, *Science Area Electives at the Massachusetts Institute of Technology* (Cambridge, MA: MIT Press, 1964); Harold S. Mickley et al., *Changing the Undergraduate Curriculum at M.I.T.: Report of the*

Committee on Educational Policy (Cambridge, MA: MIT Press, 1964); Hoffman et al., *Creative Renewal*, chap. 2. Zacharias chaired the Committee on Curriculum Content Planning in 1964. The committee's acronym, "CCCP," no doubt inspired a few chuckles. A decade earlier, Zacharias had been unfairly implicated in an alleged Communist plot to stall development of the hydrogen bomb—a scandal that became public during J. Robert Oppenheimer's infamous security hearing in 1954. The baseless charges against Zacharias went nowhere, and Zacharias went on to serve on the President's Science Advisory Committee, among other honors. By adopting the same acronym for his curriculum committee as the common abbreviation for the Soviet Union—CCCP comes from a transliteration of the Cyrillic initials of Union of Soviet Socialist Republics—Zacharias got the last laugh. See Patricia McMillan, *The Ruin of J. Robert Oppenheimer and the Birth of the Modern Arms Race* (New York: Penguin, 2005), 163, 219; John Rudolph, *Scientists in the Classroom: The Cold War Reconstruction of American Science Education* (New York: Palgrave, 2002); Jack S. Goldstein, *A Different Sort of Time: The Life of Jerrold R. Zacharias, Scientist, Engineer, Educator* (Cambridge, MA: MIT Press, 1992).

25. Lewis et al., *Report*, 21, 105; see also 26–27, 42.

26. Lewis et al., *Report*, 42, 43–44, 45–46. Similar sentiments were expressed in Robert G. Caldwell, "The Four-Year Program," in *Humanities and Social Sciences at the Massachusetts Institute of Technology* (Cambridge, MA: MIT Press, 1947), 1–3. The pamphlet is available in T171.M42g.H86 1947, MIT Institute Archives and Special Collections, Cambridge, Massachusetts.

27. Allan A. Needell, "Project Troy and the Cold War Annexation of the Social Sciences," in *Universities and Empire: Money and Politics in the Social Sciences during the Cold War*, ed. Christopher Simpson (New York: New Press, 1998), 3–38; Allan A. Needell, "'Truth Is Our Weapon': Project Troy, Political Warfare, and Government-Academic Relations in the National Security State," *Diplomatic History* 17 (June 2007): 399–420. See also Donald M. Blackmer, *The MIT Center for International Studies: The Founding Years, 1951–1969* (Cambridge, MA: MIT Center for International Studies, 2002), chap. 1; Kenneth Osgood, *Total Cold War: Eisenhower's Secret Propaganda Battle at Home and Abroad* (Lawrence: University Press of Kansas, 2006).

28. Nelkin, *The University*; Leslie, *The Cold War*, chap. 9; Geiger, *Research and Relevant Knowledge*, 191–194, 241–245; Daniel Kevles, *The Physicists: The History of a Scientific Community in Modern America*, 3rd ed. (1978; repr., Cambridge, MA: Harvard University Press, 1995), chaps. 24–25.

29. Hoffman et al., *Creative Renewal*, 1, 43.

30. Lewis et al., *Report*, 16, 60.

31. Hoffman et al., *Creative Renewal*, 66.

32. Ibid., 12–13, 16–17, 22, 209–219. For data on undergraduate subject enrollments, see Stuart, *Facts about MIT*, 123–131.

33. Hoffman et al., *Creative Renewal*, 1, 78–82.

34. Ibid., 83–84.

STUART W. LESLIE

The Vietnam War engaged MIT much as it did other U.S. universities but with one key difference. On other campuses, the war sparked protests and even riots intended to call attention to what radicals deemed misguided national policies. MIT's students and faculty instead challenged the administration *and themselves* to reexamine their own priorities in light of questions about the social responsibility of scientists and engineers as well as MIT's proper role "in the nation's service."[1] At the center of this debate was the question of what to do with the so-called special laboratories—the Lincoln Laboratory and the Instrumentation Laboratory—cold war legacies that had grown to be as large as MIT itself. What set them apart from other MIT laboratories were their sheer size, large professional staffs, strongly mission-oriented research, almost total dependence on contract research, and large number of classified projects. Did they contribute appropriately to MIT's educational and research programs by providing unmatched opportunities for real-world engineering experience, as their defenders insisted? Or had they become too dependent on military funding (much of it for classified research) and too independent of the university's administrative control to remain part of an academic community, as their critics argued? At what point did the kind of public service taken for granted during the cold war undermine MIT's larger purpose as an institution of higher, rather than "hire," learning?

As distinguished linguist and leading faculty activist Noam Chomsky forcefully pointed out, MIT could not sidestep the bigger issue: "In an institution largely devoted to science and technology, we do not enjoy the luxury of refusing to take a stand on the essentially political question of how science and technology will be put to use, and we have a responsibility to take our stand with consideration and care."[2] The debate about the military funding of academic science and engineering, and the place of the special laboratories at MIT, would in fact be undertaken with deliberation, although not without occasional rancor. The ultimate decision to divest the Instrumentation Laboratory and retain the Lincoln Laboratory as a federally funded research and development center managed by MIT seemed decisive at the time. In retrospect, less changed than either side in the debate envisioned.

PROTEST AND RESPONSE ON CAMPUS

Political cartoonist Paul Conrad perfectly captured the essential paradox of radicalism at MIT in a panel featuring a scruffy student carrying a placard filled with equations over the caption, "Man, that's really telling it like it is!" which *Technology Review* reprinted in 1969. There had been scattered demonstrations against CIA funding for the Center for International Studies in 1966, a sit-in led by the Students for a Democratic Society against Dow's campus recruiting in 1967 to protest the company's manufacture of Napalm, and a weeklong "sanctuary" for an Army deserter in 1968. Compared to Berkeley, Columbia, or Harvard, however, where students fought pitched battles against the police, MIT had been relatively quiet. Indeed, the faculty had been more keenly attuned to the arms race, military funding of academic research, and the escalating war in Vietnam than the students.

Prominent MIT faculty members, many of them Manhattan Project and Rad Lab veterans, had been outspoken opponents of nuclear weapons research, development, and testing, early antiballistic missile proposals, and the Vietnam War. Back in 1946, mathematician Norbert Wiener had famously declared that he would no longer work on or publish research "which might do damage in the hands of irresponsible militarists."[3] And a decade before students put conversion of the special laboratories on the agenda, Louis Smullin, a longtime member of the Research Laboratory of Electronics, proposed a civilian version of the Lincoln Laboratory to study "the major nonmilitary engineering problems of the modern world," including alternative fuels, soil conservation, the desalination of seawater, and other challenges facing the developing and developed world alike.[4]

Yet however sympathetic Smullin and his colleagues might have been to student demands for personal and institutional accountability, they pre-

FIGURE 6. I
When Army private J. Michael O'Connor deserted his assigned post in fall 1968, MIT students offered him sanctuary on campus. A round-the-clock vigil protected him from arrest for a full week during late October and early November 1968.
Source: Courtesy of the MIT Museum.

ferred working within the system, often as highly placed advisers to government agencies, including the military. Physicist Jerrold Zacharias bragged about his "seven hundred days in Washington" and colorfully explained, "What these guys [the student protesters] haven't figured out is the most basic axiom of politics, namely, 'It's better to be on the inside heaving out than on the outside heaving in!'" Jerome Wiesner, professor, provost, science adviser to President John F. Kennedy, and future MIT President, rightly chided antiwar protesters that he had "done more for peace than you ever had."[5] But being an insider usually meant working directly with the military-industrial complex, whether on sponsored research projects or summer studies. Zacharias and Wiesner, among other MIT faculty, had been members of the Cambridge Discussion Group, which consulted with the Department of Defense on classified technological fixes for the Vietnam

War, including the controversial electronic fence to cut off the Ho Chi Minh trail.[6]

So it came as a nationwide shock when a group of MIT faculty and students announced in January 1969 that they were calling for a research stoppage on March 4, not as a protest against MIT itself, they stressed, but rather as a symbolic gesture intended to provoke "a public discussion of problems and dangers related to the present role of science and technology in the life of our nation."[7] The idea originated with the newly formed Science Action Coordinating Committee (SACC), a group of student activists who opposed the war, academic credit for classified research, and what they considered the militarization of the academy.[8] The press could not resist dubbing March 4 a "strike," although the signers of the March 4 manifesto, including forty-eight faculty members, preferred to describe it as a "day of reflection," a time to collectively consider ways of "turning research applications away from the present emphasis on military technology toward the solution of pressing environmental and social problems."[9]

Liberal journalists applauded the organizers for "starting to behave like citizens as well as scientists," while conservatives wondered if MIT would end up sliding down a slippery slope to what would today be labeled political correctness.[10] Predictably, most MIT students and faculty spent the big day in their classrooms and laboratories. A few hard-liners among them even organized a "work-in" as a counterprotest. Still, an estimated 1,400 students (out of a total of 7,764) jammed Kresge Auditorium to hear panel discussions on conversion, arms control, and the responsibility of intellectuals. Widely covered in the press, March 4 touched off a national debate on the military's presence on campuses, classified research, and national research spending priorities. It inspired sit-ins, teach-ins, and similar events at some thirty other universities, and plunged MIT into an intensive debate about its own role as the country's largest university defense contractor and the only university among the top one hundred military contractors (it ranked fifty-fourth).

Taking a longer view, the forty-eight faculty signers of the March 4 manifesto organized themselves in 1969 as the Union of Concerned Scientists (UCS), and pledged themselves to "a critical and continuing examination of governmental policy in areas where science and technology are of actual or potential significance."[11] What began as a small coalition ultimately grew into the largest nonprofit organization of its kind, with a

FIGURE 6.2
Student activists, joined by forty-eight faculty members, organized a "day of reflection" on March 4, 1969, to assess the place of military research on campus.
Source: Courtesy of the MIT Museum.

quarter-million members worldwide. Still headquartered in Cambridge, UCS has become a powerful scientific voice on such matters as nuclear weapons and power, climate change, and alternative energy.[12]

March 4 put "war research" and "conversion" on the MIT agenda, and made the Instrumentation and Lincoln laboratories convenient targets. With annual research budgets of $55 million and $59 million, respectively, in 1969 ($324 million and $328 million in 2008 dollars), the special laboratories each spent about as much as the rest of MIT's sponsored research programs combined. Together the two laboratories represented just over half of MIT's total budget.[13] Not only were the special laboratories funded almost entirely by federal contracts (the Instrumentation Laboratory's budget was roughly split between the Department of Defense and NASA; the Lincoln Laboratory's budget came completely from the Department of Defense), those contracts supported projects with direct military applications. The Instrumentation Laboratory specialized in inertial guidance systems for intercontinental ballistic missiles (including the latest generation

Poseidon submarine-based missile), although it also had responsibility for developing the guidance system for the Apollo moon program—a perfect example of dual-use technology. The Lincoln Laboratory, founded and funded by the Air Force to develop a continental air-defense system, specialized in advanced radar, space communications, and ballistic missile reentry systems. Aside from the Lincoln Laboratory's relatively small battle-field surveillance program, which used radar to identify enemy troops camouflaged by the jungle, neither laboratory contributed much to the Vietnam War directly. All the same, the laboratories were vulnerable, if only symbolically, especially the Instrumentation Laboratory, which was housed in converted factory and warehouse buildings just blocks from the main campus. The Lincoln Laboratory, nearly twenty miles west, was out of range and secure within an Air Force base.[14]

"The students could gin up a march on the steps of 77 [the Rogers Building] in the sun, put themselves in a frenzy, and be over here in two or three minutes," ruefully recalled Charles Stark Draper, the director of the Instrumentation Laboratory. On April 22, 1969, SACC organized a rally in front of Building 10 (the Maclaurin buildings) to protest the Poseidon missile research and marched on the Instrumentation Laboratory for a direct confrontation. Draper caught them by surprise, meeting them at the door and inviting some of them into the laboratory for a firsthand look at the military-industrial complex in action. After a brief verbal exchange (in which Draper said he had been able "to give them hell on their own bullhorn several times"), the protesters returned to campus to press their demands with the administration.[15]

The demonstration, mild enough compared with disorders on other campuses that spring, still alarmed an administration already edgy over recent events at Harvard and unaccustomed to student activism. Responding to an obviously energized, if not radicalized, campus community, President Howard Johnson hastily convened a twenty-two-member Review Panel on Special Laboratories, and charged it with evaluating the relationship between the laboratories and the rest of MIT, specifically "the implications that the laboratories have for the Institute in its prime responsibility for education and research and in its responsibility for service to the nation."[16] Johnson had moved up to the President's office in 1966 after eight years as Dean of the Sloan School of Management. He preferred consensus to confrontation and hoped the review would appease the protesters. Popularly dubbed the Pounds Panel, for chair William F. Pounds (Johnson's successor at the Sloan School), the panel took up its work at a furious pace—furious even by MIT standards. In little more than a month, its members met twenty times for an estimated total of 109 hours. The panel interviewed a hundred people—MIT faculty, special laboratories staff, students, and outside experts—and made side trips to consult with VIPs in Washington, DC, and California.[17] Although the special laboratories would later claim that the Pounds Panel had been stacked against them, each laboratory had two representatives. From the other end of the political spectrum, the committee included Chomsky and Jonathan Kabat, one of the founding members of SACC.

CONVERSION OR DIVESTMENT?

Achieving anything like a consensus on such a controversial subject in a matter of weeks would have been expecting too much. Nevertheless, the review panel did agree on some guiding principles. Its final recommendations included diversifying the laboratories' research portfolios to achieve a more even balance of military and civilian research, fostering more collaboration between the laboratories and the rest of the campus, reducing the level of classified research at the laboratories, and appointing a watchdog committee to monitor future laboratory contracts and policy. Chomsky appended what amounted to a lengthy minority opinion, arguing that the recommendations did not go nearly far enough in pushing the laboratories to redirect their expertise toward "socially useful technology," and he added his signature and support to an even longer and more strident minority report contributed by Kabat that called for "total conversion" of the laboratories. Only two panel members, chemical engineer Edwin Gilliland and electrical engineer Marvin Sirbu, called for prompt divestment. They were convinced that the special laboratories had contributed to a growing "imbalance between research and education" that was "changing the character of the Institute." What MIT needed, they said, were smaller, flexible, mission-oriented laboratories with limited life spans.[18]

The key question facing the Pounds Panel, and the MIT community as a whole, was whether to cut the laboratories loose and let them pursue their unique research agendas unhindered, or embrace them more closely with the aim of rechanneling their technical expertise toward nonmilitary

FIGURE 6.3
William F. Pounds, Dean of MIT's Sloan School of Management, introducing members of the Review Panel on Special Laboratories at the panel's first press conference in May 1969.
Source: Courtesy of the MIT Museum.

applications. President Johnson advocated a compromise between "complete divestment and complete conversion" that satisfied neither the activists nor the laboratories' staffs and sponsors.

The proponents of conversion drew inspiration from the example of the Fluid Mechanics Laboratory. Founded in 1962 as part of the mechanical engineering department, the laboratory initially specialized in reentry physics for missiles and spacecraft.[19] By 1966 the laboratory had six professors, twenty graduate students, and a $300,000 budget (nearly $2 million in 2008 dollars), none of it from classified contracts but virtually all from defense agencies. Worried about the boom-and-bust funding and employment cycles characteristic of the aerospace industry, the laboratory decided to seek a more balanced research program with particular attention to "socially oriented" projects, such as air and water pollution, biomedical engineering, and the desalination of seawater. In just three years, the Fluid Mechanics Laboratory built a new network of sponsors, doubled its budget (two-thirds of it from nondefense sources), and sent half of its graduates to civilian industries.[20] As laboratory director Ronald Probstein stressed in the March 4 discussions, he and his colleagues had no objection to working on defense contracts. Indeed, they praised the Office of Naval Research and the laboratory's other military sponsors for a "sensitive and intelligent approach to research." Their goal was to instead "redress an imbalance" in the current funding climate, and encourage other government agencies and the private sector to support the laboratory's research aimed at solving environmental and urban issues.[21]

Despite their own experience, and contrary to the expectation of many of the activists, the members of the Fluid Mechanics Laboratory did not support the conversion of the special laboratories. As Ascher Shapiro, head of the mechanical engineering department and a member of the Fluid Mechanics Laboratory explained, the real objective should be "converting" the national agenda: "The only way to bring the military back into reasonable balance is to attack the problem at the national level."[22] Trying to convert the special laboratories would end up being counterproductive, he said. At best, the special laboratories would end up competing for scarce civilian funding with other MIT laboratories, such as Shapiro's own.

The leaders of the special laboratories were likewise skeptical about conversion. Draper dismissed it as so much wishful thinking. "The weakness of all these gentlemen who talk [about conversion] is that they are completely devoid of ideas," he maintained. "They're not about to do anything. They're only making noises about how other people should do it. They have no practical suggestions to offer from the standpoint of financing, of organization, of subject matter."[23] And, he might have added, no understanding of the kind of budget it took to run an operation the size of the Instrumentation Laboratory. How was MIT going to raise $10 to $15 million a year for civilian technology at a time of shrinking federal budgets (roughly $60 to $90 million in 2008 dollars)? Even that amount would be only a fifth of the current Instrumentation Laboratory's total budget. "All the Alumni Organization [*sic*] together, working for a whole year, can only raise two weeks worth of funding for this Lab," its associate director William Denhard cautioned.[24]

The Pounds Panel recommendations sufficiently troubled the Instrumentation Laboratory's sponsors at the Air Force that they drew up contingency

FIGURE 6.4
Protests against military research on campus flared at Alumni Day, on June 16, 1969.
Source: Courtesy of the MIT Museum.

plans for transferring its research programs to private industry or reincorporating the laboratory as an independent, nonprofit entity along the lines of MITRE, a spin-off from the original Lincoln Laboratory.[25] Draper, whose homespun testimony had charmed the Pounds Panel, put on a resolute public face but privately called the panel an "inquisition" and hinted that if pressed too far, he would take the laboratory elsewhere, despite his forty-seven-year tenure at MIT. His lieutenants, seeing little prospect for large-scale funding aside from the Department of Defense and NASA, considered divestment inevitable and the laboratory's future in serious jeopardy. Meanwhile, Lincoln Laboratory director Milton Clauser had a slightly different view. He asserted that the activists had "shaken some of the lethargy out of us" and thought more engagement with civilian technologies might actually be good for his laboratory.[26] The Lincoln Laboratory, however, had the advantage of a long-term contract and a broad range of research interests with potential civilian applications, such as advanced radar systems for air-traffic control and solid-state electronics.

The pressure on the special laboratories only intensified when the students and faculty returned to campus in fall 1969. In September, President Johnson named the oversight committee recommended by the Pounds Panel (to be chaired by chemist John Sheehan) and announced that civil engineer Charles Miller, head of MIT's Urban Systems Laboratory, would take over as director of the Instrumentation Laboratory, with Draper as technical director and senior advisor. Draper made no attempt to disguise his feeling that Miller's appointment was a stab in the back and told the press, "I've been fired." In a major departure from past policy, Johnson also declared that MIT would no longer accept contracts for the "design and development of systems intended for operational deployment as military weapons."[27] Draper scathingly dismissed the notion that his nineteen hundred scientists and engineers would have any interest in redirecting their research. "These people are used to working on very sophisticated problems like astrodynamics," he noted. "There's no chance of any substantial number who would be attracted to do jobs on conversion problems," even if the money could be found—itself a doubtful proposition.[28]

If Johnson believed a change of leadership and policy would calm the campus, he was dead wrong. Draper urged his staff to prepare for an impending "time of troubles." Emboldened by their apparent success, radical students pushed for a more rapid reordering of MIT's priorities and started talking about shutting down the "war machine" by force. Taking the threat seriously, Johnson took the unprecedented step of obtaining a restraining order prohibiting demonstrators from the use or even threatened use of force. Undeterred, some 350 students marched on the Instrumentation Laboratory on November 5, waving Viet Cong flags and shouting "shut it down!" at the technicians inside. They scuffled with a few of the laboratory workers arriving for the morning shift. A short time later, city police moved in with dogs and tear gas at the ready, and promptly routed the protesters down the back alleys of Cambridge. Within an hour, the laboratory reopened for business.[29] Even though only ten people had been injured and one arrested, the "siege" at the Instrumentation Laboratory provoked widespread comment and debate. The policy of conversion and Draper's impending "retirement" signaled to conservatives a pattern of appeasement to student radicals, and led like-minded pundits to call for putting the laboratory under direct federal control. To liberal observers, the

FIGURE 6.5
Antiwar protesters marched to MIT's Instrumentation Laboratory on the morning
of November 5, 1969, objecting to the laboratory's heavy involvement with military research.
Source: Courtesy of the MIT Museum.

laboratory's heavy dependence on military funding provided an instructive
example of misplaced academic priorities.[30]

TOO ENTWINED TO PULL APART

One of the chief recommendations of the Pounds Panel had been strengthening ties between the special laboratories and the rest of campus. Yet the
Pounds Panel's own study revealed how much the two laboratories already
contributed to MIT's teaching and research. The Instrumentation Laboratory was officially part of the Department of Aeronautics and Astronautics
("aero and astro"), and 16 faculty members in the department held appointments at the laboratory. In 1969, 398 MIT students had some affiliation
with the laboratory, including 105 graduate students in aero and astro (half
of the entire department's total enrollment), 159 students from electrical
engineering, and smaller but still significant numbers from mechanical

FIGURE 6.6
Riot police clashed with antiwar protesters outside MIT's Instrumentation Labora-
tory on November 5, 1969.
Source: United Press International photograph, courtesy of Corbis.

engineering, physics, mathematics, and even management, plus 200 full-
time summer students. Add to that 25 courses taught in aero and astro (out
of 80 offered that year) on subjects directly related to Instrumentation
Laboratory specialties, with many of those courses taught by lab staff, plus
the 36 completed master's and doctoral dissertations based on research at
the laboratory, and the full impact became clear.[31]

Moreover, the Instrumentation Laboratory offered the kind of hands-on
engineering nearly impossible to find in its contemporary academic coun-
terparts. The lab reached back to an older MIT tradition (in which Draper
and many of his project leaders had been steeped) that did not so clearly
distinguish engineering theory from practice, and whose ultimate measure
of success was in the field or on the factory floor. As Draper emphasized
in his testimony to the Pounds Panel, in a field like inertial guidance,
engineering practice generally outpaced engineering theory. Students at the
Instrumentation Laboratory learned what no textbook could teach them,
and no one there could cheat on the final exam. "Their equipment is riding
the bird [flying aboard an actual missile]," he explained, "and if it doesn't

work, why, you know it as well as everybody else, and they know it, too."[32] At the Instrumentation Laboratory, students learned some invaluable skills, including meeting deadlines, keeping to the bottom line, and dealing with industrial contractors, union machinists, and government sponsors. Draper prided himself on his teaching, and always considered the Instrumentation Laboratory a classroom and instructional laboratory—something very different from MITRE, RAND, or Battelle. Rene H. Miller, who was Draper's successor as department head of aero and astro, compared the Instrumentation Laboratory to a teaching hospital at a top medical school, a place where engineering students could gain experience that "can only be acquired through deep involvement in the agonizing daily decision-making process of a large engineering enterprise." The postwar MIT curriculum, he said, with its stronger emphasis on "engineering science" (mathematics, physics, and theory), had pushed such real-world experience to the margins, to the peril of future engineers.[33]

The Lincoln Laboratory had less direct contact with the campus—hardly surprising given its distant location. Still, thirty students a year wrote doctoral dissertations on research done there, a similar number spent one semester during the academic year working full-time in the lab, and twenty-three staff members had appointments on the MIT faculty. Both laboratories, of course, were heavily staffed with MIT graduates. Just under a third of the Lincoln Laboratory's professional staff had MIT degrees.[34]

On May 20, 1970, after a year under the oversight committee, the special laboratories learned their fate. President Johnson told the faculty that MIT would divest the Instrumentation Laboratory but retain the Lincoln Laboratory. Johnson had decided the matter much earlier, telling the incoming director of the Instrumentation Laboratory, Charles Miller, about divestment back in January. Despite its closer and denser connections to campus, and a far better balanced research program, the Instrumentation Laboratory got the ax. It was too visible and too close to home. The Lincoln Laboratory would remain a federally funded research and development center, and be encouraged to seek support beyond military agencies.

Draper and his colleagues took the news hard as well as personally. Deputy Assistant Director of the Instrumentation Laboratory Charles Broxmeyer had bitterly and prophetically told the Pounds Panel that what MIT seemed to be looking for was a scapegoat: "The special laboratories, which every shred of evidence indicates have nothing whatever to do [with the genesis] of the Vietnam War, have been selected to be destroyed. Only then will the collective guilt of the MIT community, that part of it which remains, be expiated."[35] The decision also disappointed the proponents of conversion, such as Chomsky, who had hoped that the special laboratories might become a test bed for redirecting America's scientific and engineering resources in nonmilitary directions. On June 1, 1970, the Instrumentation Laboratory (now renamed the Charles Stark Draper Laboratory for its founder) became an independent division of MIT under its own board. After a three-year transition period to sort out contractual obligations, it became an independent nonprofit research corporation (with no formal contractual ties to MIT); John Duffy served as its first president.[36] Charles Miller stepped down from the Draper Laboratory to direct the Urban Systems Laboratory that he had founded at MIT in 1968 with support of the Ford Foundation. Designed as an interdisciplinary effort to apply engineering techniques to the problems of the city, Urban Systems quickly ran out of money and closed its doors in 1974.[37] Draper and the other skeptics of conversion had been right about the difficulty in finding contracts to support large-scale programs in civilian technology.

The special laboratories prospered in their new roles, though. Draper Laboratory built a handsome new headquarters in Technology Square in Cambridge and moved there in 1977. As the critics predicted, divestment only strengthened Draper Laboratory's dependence on defense contracts. It instantly vaulted to the top of the nonprofit federal contractor list with $83 million in funding for 1974 ($360 million in 2008 dollars). Freed from the restrictions of the oversight committee, Draper Laboratory gave full attention to its ballistic missile guidance systems for the Trident and Peacekeeper programs. Building on its Apollo success, it also developed the guidance system for the space shuttle.[38] Meanwhile, the Lincoln Laboratory made a good faith effort to add nondefense contracts and had some success. It won its first civilian contract in 1971, and broadened its portfolio to include sponsored research on air-traffic control, solar energy, and health care, although defense contracts consistently accounted for 80 to 90 percent of its annual budget.[39]

With the end of the Vietnam War, and the beginning of the "war on cancer" and other health initiatives, national funding patterns changed significantly. Department of Defense–sponsored research dipped both absolutely and as a fraction of the total MIT research budget, falling by half from 1967 to 1977. By 1978, the Department of Energy had become MIT's biggest sponsor, followed closely by Health and Human Services (primarily the National Institutes of Health) and the National Science Foundation; the Department of Defense dropped to fourth. MIT nonetheless remained atop the list of the nation's leading university defense contractors, even after the divestment of the Instrumentation Laboratory and excluding the Lincoln Laboratory's budget. From 1974 to 1984, the Institute ranked first in all but two years (when it dropped slightly behind Penn State), averaging $17 million per year in Department of Defense funding (roughly $55 million in 2008 dollars).[40]

The debate over the Strategic Defense Initiative (SDI) in the 1980s revealed the lengthening shadow cast by the special laboratories debate. In 1985 a group of faculty members, apprehensive about the long-term consequences of the SDI, met with MIT President Paul Gray and subsequently organized itself as the Ad Hoc Committee on the Military Presence at MIT. Chaired by political scientist Carl Kaysen, the committee surveyed MIT faculty and students to gauge their response to the Reagan-era military buildup and the SDI contracts. A surprisingly high response rate—45 percent of the faculty, 17 percent of the graduate students, and 20 percent of the undergraduates participated in the survey—suggested genuine interest in the subject. Perhaps not coincidentally, the committee asked the Office of Sponsored Programs for a list of SDI-funded projects on March 4, 1985— the sixteenth anniversary of the day of reflection in 1969. The survey results indicated widespread concern about the SDI's potential impact on MIT, starting at the top. President Gray himself addressed the matter at commencement that spring: "What I find particularly troublesome about the SDI funding is the effort to short-circuit the debate and use MIT and other universities as political instruments in the attempt to obtain implicit institutional endorsement. This university will not be so used."[41] A majority of the faculty agreed with Gray's further statements: "SDI funding should be avoided because it may impose changes in priorities in educational and research programs at MIT," and "SDI funding should be avoided because MIT funding should not be involved in weapons system development." And they strongly opposed classified research and security clearances for graduate students as antithetical to the tradition of academic freedom. Students did not take to the streets as they had in 1969, although significant numbers of graduates in the class of 1985 (45 percent in electrical engineering and physics, 27 percent in mechanical engineering, and 21 percent in aero and astro) said they "felt strongly against working in defense."[42]

Despite this critical, if not hostile, climate, the special laboratories remained bastions of loyal opposition with considerable sway over the rest of MIT. The ad hoc committee concluded that even with divestment, Draper Laboratory's "campus influence remains essentially the same. Whether measured at the faculty-staff level or in terms of opportunities for student research support, Draper continues to have a greater influence on the MIT student environment than does Lincoln."[43] And the Lincoln Laboratory had as much collaboration with the main campus as it ever had: through faculty appointments, research assistantships, and consulting arrangements. All that had changed in two decades was that the Lincoln Laboratory had a bigger budget ($386 million in 1988, or $700 million in 2008 dollars) and a higher percentage of military funding. If anything, the special laboratories provided MIT with a cordon sanitaire, a haven where faculty and students could pursue classified research beyond the reach of campus restrictions. Conversion had indeed been a chimera.

Nowadays, MIT and its peer institutions must wrestle with yet another round of classified contracts, this time in support of the war on terrorism. The subspecialties may have changed—to surveillance, computer and information security, cryptography, and potentially at least, (anti)biological warfare—but not the fundamental issues raised by the original special laboratories. MIT's administration has spoken forcefully on behalf of openness and access. A panel that was convened to reconsider policy in the wake of September 11, 2001, recommended closer engagement with the Lincoln and Draper laboratories to "provide facilities for faculty to carry out classified research in compatible areas."[44] The panel raised the possibility of a laboratory for classified biological research, modeled on the Lincoln Laboratory—although the panel also reaffirmed MIT's policy prohibiting classified research on campus.

The special laboratories may be part of MIT's institutional DNA, but that inheritance also includes a continuing and healthy debate about MIT's stated mission of "service to the nation and service to humanity." Just as the special laboratories have not changed all that much over the past forty years, neither has the spirit of critical inquiry that gave birth to the Union of Concerned Scientists. Learning how to make informed decisions has been, and must be, the hallmark of a socially responsible technical education.

NOTES

1. For the antiwar movement on campus, see Kenneth Heineman, *Campus Wars: The Peace Movement at American State Universities in the Vietnam Era* (New York: New York University Press, 1993); Jerry Avorn, *Up against the Ivy Wall* (New York: Library Press, 1969); Cox Commission, *Crisis at Columbia: Report of the Fact-Finding Commission Appointed to Investigate the Disturbances at Columbia University in April and May 1968* (New York: Random House, 1968); Tom Bates, *Rads: The 1970 Bombing of the Army Math Research Center at the University of Wisconsin and Its Aftermath* (New York: HarperCollins, 1992).

2. Noam Chomsky, quoted in Review Panel on Special Laboratories, *Final Report* (October 1969), 32. Available in Activism Reports, MIT Museum, Charles Stark Draper Laboratory Historical Collection, Cambridge, Massachusetts.

3. Norbert Wiener, "A Scientist Rebels," *Atlantic Monthly* 179 (January 1947): 41. For Wiener's principled opposition, see Steven Heims, *John Von Neumann and Norbert Wiener: From Mathematics to the Technologies of Life and Death* (Cambridge, MA: MIT Press, 1980).

4. Louis Smullin, "Proposal for a Study of Major Engineering Problems of the World," August 26, 1959, AC 134/21, "Electrical Engineering," MIT Institute Archives and Special Collections, Cambridge, Massachusetts.

5. Jerome Wiesner, quoted in Richard Todd, "The 'Ins' and 'Outs' at M.I.T.," *New York Times Magazine*, May 18, 1969, 32.

6. For details on the contributions of the Cambridge Discussion Group, see Ann Finkbeiner, *The Jasons: The Secret History of Science's Postwar Elite* (New York: Viking, 2006), 65–82.

7. Boris Magasanik, John Ross, and Victor Weisskopf, "No Research Strike at M.I.T.," *Science* 163 (February 7, 1969): 517. For an indispensable source on the war research debate at MIT, see Dorothy Nelkin, *The University and Military Research: Moral Politics at MIT* (Ithaca, NY: Cornell University Press, 1972).

8. For March 4 and its aftermath, see Kelly Moore, *Disrupting Science: Social Movements, American Scientists, and the Politics of the Military, 1945–1975* (Princeton, NJ: Princeton University Press, 2008), 137–146.

9. "The Misuse of Science," *Nation*, February 24, 1969, 228.

10. Ibid.

11. The founding document of the Union of Concerned Scientists is available at <http://www.ucsusa.org/about/founding-document-1968.html> (accessed August 7, 2008).

12. For information on the Union of Concerned Scientists, see <http://www .ucsusa.org> (accessed December 8, 2008).

13. Nelkin, *The University*, 18.

14. For the official history of the Lincoln Laboratory, see Eva Freeman, ed., *MIT Lincoln Laboratory: Technology in the National Interest* (Lexington, MA: Lincoln Laboratory, 1995). On the Instrumentation Laboratory's contributions to NASA's Apollo program, see also David Mindell, *Digital Apollo: Human and Machine in Spaceflight* (Cambridge, MA: MIT Press, 2008).

15. Charles Stark Draper, oral history interview with Barton Hacker, January 19, 1976, 120. Transcript available in MC134 (1), MIT Institute Archives and Special Collections, Cambridge, Massachusetts.

16. Review Panel on Special Laboratories, *Final Report*, 6.

17. Ibid., 149.

18. Ibid., 44–45.

19. For the Fluid Mechanics Laboratory in the larger context of engineers as activists, see Matt Wisnioski, "Inside 'The System': Engineers, Scientists, and the Boundaries of Social Protest in the Long 1960s," *History and Technology* 19 (2003): 313–333.

20. Peter Guynne, "A Physics Lab Goes Relevant," *Science News* 96 (August 16, 1969): 132.

21. Wisnioski, "Inside 'The System,'" 322.

22. Ascher Shapiro, "A Position Paper on Retention of Divestiture of the Special Laboratories," February 1970, Activism Reports, MIT Museum, Charles Stark Draper Laboratory Historical Collection, Cambridge, Massachusetts.

23. Charles Stark Draper, quoted in William Leavitt, "The Dethronement of Dr. Draper," *Air Force/Space Digest* (December 1969): 49.

24. William Denhard to Jerome Wiesner, October 31, 1969, Activism Correspondence, MIT Museum, Charles Stark Draper Historical Collection, Cambridge, Massachusetts.

25. David Dyer and Michael A. Dennis, *Architects of Information Advantage: The MITRE Corporation since 1958* (Montgomery, AL: Community Communications, 1998).

26. Milton Clauser, quoted in "Can Defense Work Keep a Home on Campus?" *Business Week*, June 7, 1969, 68.

27. Howard Johnson and Charles Stark Draper, quoted in "Can a Weapons Lab Solve Urban Ills?" *Business Week*, November 1, 1969, 132.

28. Charles Stark Draper, quoted in "Go Back? Go Back!" *Newsweek*, November 17, 1969, 79.

29. "Treading the Narrow Line: The First Week of November," *Technology Review* 72, no. 6 (December 1969): 96B.

30. For the opposing positions, see "New Left v. National Security," *Nation*, January 13, 1970, 18; "Day of Reckoning," *Nation*, November 24, 1969, 556.

31. Review Panel on Special Laboratories, *Final Report*, 63–64.

32. Charles Stark Draper, testimony to the Review Panel on Special Laboratories, *Final Report*, 58.

33. Rene H. Miller, "The Draper Laboratory and the Department of Aeronautics and Astronautics," February 24, 1970, Activism Reports, MIT Museum, Charles Stark Draper Laboratory Historical Collection, Cambridge, Massachusetts.

34. Review Panel on Special Laboratories, *Final Report*, 69.

35. Charles Broxmeyer, quoted in Leavitt, "The Dethronement," 50.

36. For an account of the divestment from the perspective of the Instrumentation Laboratory, see Christopher Morgan with Joseph O'Connor and David Hoag, *Draper at 25: Innovation for the 21st Century* (Cambridge, MA: Draper Laboratory, 1998), 35–40, available at <http://www.draper.com/draper25/draper25.pdf> (accessed August 11, 2008).

37. For the Urban Systems Laboratory, see <http://libraries.mit.edu/archives/research/collections/collections-ac/ac366.html#history> (accessed February 27, 2009).

38. Morgan with O'Connor and Hoag, *Draper at 25*, 43–48.

39. Freeman, *MIT Lincoln Laboratory*, xv.

40. Ad Hoc Committee on the Military Presence at MIT, *Report* (April 1986), appendix C, 30, AC150, MIT Institute Archives and Special Collections, Cambridge, Massachusetts.

41. Paul Gray, quoted in ibid., 26.

42. Daniel Glenn, "MIT Research Heavily Dependent on Defense Department Funding," *The Tech* 109 (February 28, 1989).

43. Ad Hoc Committee on the Military Presence at MIT, *Report*, 16.

44. "MIT Panel Urges Off-campus Sites for Classified Research," *MIT Tech Talk*, June 12, 2002.

7 "REFRAIN FROM USING THE ALPHABET": HOW COMMUNITY OUTREACH CATALYZED THE LIFE SCIENCES AT MIT

JOHN DURANT

On April 17, 1974, a group of influential biologists met in the office of David Baltimore, a young faculty member who had recently moved into MIT's new Center for Cancer Research. The meeting had been called by Stanford biologist Paul Berg. As he explained to his colleagues in advance, he'd been invited by the National Academy of Sciences to head a study panel to "consider whether or not there is a serious problem growing out of present and projected experiments involving the construction of hybrid DNA molecules in vitro. If a problem exists, then what can be done about it, both short- and long-term actions."[1] Besides Baltimore and Berg, just six others attended the meeting: Daniel Nathans from Johns Hopkins University, James Watson from Cold Spring Harbor Laboratory, Sherman Weissman from Yale University, Norton Zinder from Rockefeller University, Richard Roblin from Harvard University Medical School, and Herman Lewis, a staff member from the National Science Foundation.

The meeting lasted all day, and by the time it broke up, a decision had been taken that contributed over the space of just a few years to the transformation of the life sciences in the United States and beyond, the launching of modern biotechnology, and the radical redefinition of the relationship between science and the wider community in the last quarter of the twentieth century. At MIT, the growth of the life sciences and technology eventually changed the Institute in ways that few of those who gathered that day in Baltimore's office could possibly have foreseen.

A LETTER AND A MONKEY WRENCH

The kinds of research that worried Berg and his colleagues involved recombining DNA from different organisms to create hybrids with entirely novel genetic characteristics. Several apparently unrelated discoveries about the genetics of bacterial and animal cells had opened up a number of surprisingly straightforward ways to do this. For example, bacteria had been found to contain small loops of extrachromosomal DNA, known as plasmids, which could be exchanged between cells. In addition, a special class of "restriction enzymes" had been discovered that were capable of cutting DNA into fragments with complementary or "sticky" ends. Add a restriction enzyme to two different DNA samples taken from two completely different sources (say, a bacterial plasmid and a mouse cell), mix the resulting fragments together, and with a bit of luck, the sticky fragments would recombine in ways that brought together bacterial and mouse genes in a single "recombinant molecule," ready for transfer back into living cells.

In the early 1970s, recombinant DNA (rDNA) techniques were full of promise for fundamental research, but they also raised some worrisome possibilities. What if genes conferring resistance to specific antibiotics were recombined into naturally pathogenic bacteria? Or what if genes causing cancer were transferred from viruses into the normally harmless bacteria that inhabit the human gut? Scenarios like these had been discussed at the Gordon Conference on Nucleic Acids in New Hampton, New Hampshire, in summer 1973, and conference members had then voted to send a letter of concern to *Science* magazine.[2] It was this letter that prompted the National Academy of Sciences to ask Berg to convene a panel to investigate the whole matter.

So what happened at the meeting in Baltimore's office in April 1974? Later that same year, he described it this way:

We met here at MIT in April of this year. And we sat around for the day and said, "How bad does the situation look?" And the answer that most of us came up with was that . . . just the simple scenarios that you could write down on paper were frightening enough that, for certain kinds of limited experiments using this technology, we didn't want to see them done at all.[3]

Everyone present concluded that many kinds of experiments should not be undertaken until the scientific community had a chance to consider carefully all the relevant safety issues. They also discussed the contents of an open letter that they intended to write to their scientific colleagues. Immediately after the meeting, Roblin, the young microbiologist from Harvard Medical School who had a keen interest in the ethical and social implica-

tions of recent developments in molecular biology, completed a draft of
the letter. Over the next few weeks, this was circulated within the group
and to a number of other biologists. The National Academy of Sciences
conferred formal status on the group, designating it as a committee of its
Assembly of Life Sciences. Finally, the agreed-on version of the open letter
was signed by eleven scientists and appeared in three publications—*Proceedings of the National Academy of Sciences*, *Nature*, and *Science*—in summer 1974.
Written in plain language and preceded by an international press conference
led by Berg, Baltimore, and Roblin, the committee's letter called for a
voluntary moratorium on designated categories of rDNA experiments,
urged the establishment of an experimental program at a secure facility to
assess potential hazards, and proposed a meeting to evaluate the situation
and make formal recommendations to the National Institutes of Health
(NIH) in spring 1975.[4]

The Berg letter (as it became known) had an immediate impact in the
United States and around the world. As a direct result of its publication,
the voluntary moratorium was widely instituted within the scientific community, and an international conference on rDNA was convened at Asilomar, California, in February 1975. This pivotal meeting extended the
moratorium and set out detailed recommendations to the NIH for a regulatory framework to govern this area of research in the future. According to
the Asilomar proposals, various types of rDNA research were to be classified
into levels of manageable risk, and appropriate levels of physical (P1 to P4)
and biological containment (e.g., the use of multiply disabled bacteria that
could not survive outside the laboratory) were to be attached to each level.
The NIH responded swiftly by publishing draft guidelines for rDNA
research, and analogous agencies in the United Kingdom and several other
countries set up similar arrangements. This was pretty much what Berg,
Baltimore, and their colleagues had hoped for.

Many of the scientists who became involved in the rDNA debate in the
1970s worked with a heightened sense of social responsibility that had
everything to do with their experiences of the movements and political
battles of the late 1960s, especially the role of science and technology during
the Vietnam War. This was certainly true of many of the skeptical scientists,
including faculty members at both MIT and Harvard, who led the charge
against rDNA technology.[5] But it was equally true of many advocates of
rDNA technology: most of them wanted the research to proceed within a

socially responsible regulatory framework. Baltimore belonged to the latter group. On more than one occasion, he testified that his attitudes toward rDNA were heavily influenced by his experiences as a young scientist in the late 1960s:

I guess I had kind of undergone a certain transformation over the period from probably sixty-eight to seventy, which a lot of people did, from being involved in larger political issues (as a speaker I'd been kind of involved in the left wing in San Diego, and here, to a certain extent, I'd been involved in the March 4th organiza- tion, that kind of thing) to the feeling that if I was going to do anything, it ought to be within the field I knew best because I'd been damned ineffective outside of it—like everybody else or almost everybody else. And so I was sensitized to issues that involved the biological community and felt that if I was going to put in politi- cal time, it should be in there rather than anywhere else.[6]

Once the moratorium on certain categories of rDNA experiments was instituted in 1974, Baltimore and some of his MIT colleagues found them- selves in a frustrating but unavoidable bind. They had just moved into superb new laboratories. Salvador Luria, a Nobel Laureate and one of the most formidable molecular biologists of the postwar period, had applied for and, in 1972, won a major grant from the National Cancer Institute of the NIH. The grant was for a new facility at MIT devoted to basic research on the cellular and molecular mechanisms underlying cancer. The funding was part of President Richard Nixon's "war on cancer." When this initia- tive was announced, MIT's Department of Biology was ideally placed to make the most of it: the department had been developing its strengths in cellular and molecular biology for some years.

Grant in hand, Luria and Baltimore wasted no time in creating their new cancer research facility. Suitable space was found in the Seely Mudd Build- ing (E17), a former chocolate factory on Ames Street, and new state-of- the-art laboratories were constructed to support all of the relevant research methods, including work with animal cells and animal tumor viruses in culture. In early 1974, new faculty began to arrive: from Cold Spring Harbor Laboratory came Nancy Hopkins and Phillip Sharp, from the Ontario Cancer Institute in Toronto came David Housman, and from MIT's own Department of Biology came Robert Weinberg. By summer 1974, one of the most powerful teams of cancer biologists in the world was assembled at the new Center for Cancer Research, and they were all set

FIGURE 7.1

MIT biologist and Nobel Laureate Salvador Luria (left) chats with colleagues Nancy Hopkins and David Baltimore in MIT's new Center for Cancer Research, January 1974.

Source: Courtesy of the MIT Museum.

to go with a research program that depended crucially on rDNA technology. Then the Berg letter threw the proverbial monkey wrench into the works. Sharp recently described to me the predicament he faced as "a young dude" raring to get to work in his new academic position at MIT:

When I came to MIT, one of the things I brought with me in my ice bucket was a bunch of restriction enzymes . . . because at Cold Spring Harbor I had already . . . made some major discoveries with that technology, and we were poised and ready immediately to use that technology to clone parts of viral genomes so that we could understand their function and test them in transformation assays, and analyze bigger and bigger viral genomes. David [Baltimore] particularly wanted to look at retroviral genomes, which were very hard to handle. They're not produced in large amounts because of the reverse transcriptase process. So we were all sitting here ready to do it, and then . . . David . . . initiated the letter, which led to the moratorium, which shut us down.[7]

The Berg letter and the Asilomar conference imposed a moratorium of almost two years on the work of Sharp and his colleagues. But that wasn't all. From the point of view of this nascent MIT research community, things soon got a whole lot worse.

"REFRAIN FROM USING THE ALPHABET"

The authors of the Berg letter appear to have imagined that they were initiating a professional discussion among scientists and science policymakers. But in 1975 and 1976, an increasing number of voices outside the science and policy communities joined the debate about rDNA technology. Before long, there was a full-blown public controversy over wider health, environmental, and defense issues as well as the role of citizens and their representatives in helping to determine what sorts of rDNA research should be done. Nowhere were the issues more keenly contested than in Cambridge, Massachusetts, where, ironically, MIT found itself embroiled in a public and political argument that originated in a decision by its neighbor Harvard University to get in on rDNA research.[8]

At noon on Wednesday, June 23, 1976, the U.S. Department of Health, Education, and Welfare announced the release of the long-awaited final NIH guidelines for the conduct of rDNA experiments: "The NIH guidelines will, effective today, govern research at laboratories of the NIH and those of its grantees and contractors. They are also expected to be adopted by other laboratories throughout the United States and foreign countries."[9] That very same evening a special Hearing on Recombinant DNA Experimentation was convened in Cambridge City Hall by Mayor Alfred Vellucci, who opened the proceedings with these words:

The subject matter before the Council tonight is important to all of us. No one person or group has a monopoly on the interests at stake. Whether this research takes place here or elsewhere, whether it produces good or evil, all of us stand to be affected by the outcome. As such, the debate must take place in the public forum with you, the public, taking a major role. I thank you for your interest and cooperation.[10]

These two events—the public announcement of the NIH guidelines and the public meeting in Cambridge—concerned exactly the same issue, but their tone could not have been more different. Indeed, fast-moving observers who managed to attend both meetings might have been forgiven for supposing that they were witnessing events in two parallel yet separate universes. In Washington, DC, there was an atmosphere of quiet confidence verging on self-satisfaction for a regulatory job well done. While conceding that not everyone was happy with the guidelines, the *Washington Post* nonetheless felt able to pronounce confidently that "the US scientists who urged this breakthrough—and worked hard to achieve it—deserve our gratitude and congratulations."[11] Cambridge Mayor Vellucci, however, took an opposite stance. He had called the hearing because he had heard reports of a recent open meeting at Harvard University to discuss a proposal to construct a moderate-risk (P3) containment facility for rDNA research.[12] Vellucci's reaction was swift and to the point: "They may come up with a disease that can't be cured—even a monster," he opined. "Is this the answer to Dr. Frankenstein's dream?"[13] Presiding over his first public hearing on the subject, Vellucci made clear to the leader of the Harvard delegation that had come to give evidence the terms on which he intended to conduct his investigation on behalf of the citizens of Cambridge:

Then, for the person who is speaking, kindly give your name, your address, your title, and the organization that you represent. Refrain from using the alphabet. Most of us in this room, including myself, are lay people. We don't understand your alphabet, so you will spell it out for us so we know exactly what you are talking about because we are here to listen. Thank you.[14]

Richard Leahy, Harvard's Associate Dean of Faculty, then introduced a series of witnesses on behalf of his university, but after one or two speakers had given evidence, Vellucci decided that

for the benefit of all the members of the City Council, I would like to inject this statement of questions, not to be answered at this time, but for the benefit of members of this city council who may want to ask these questions.

One. Did anyone of this group bother at any time to write to the mayor and city council to inform us you intended to carry out these experiments in the city of Cambridge, and you just said that you had public hearings.
You plan to use E. coli in your experiments. Do I have E. coli inside my body right now? That's a question. Don't answer [now], but you may, as you go along.
Does everyone in this room have E. coli inside their bodies right now?

Can you make an absolute, one hundred percent guarantee that there is no possible risk which might arise from this experimentation? Is there zero risk of danger? Answer that question, later, too, please.

Would recombinant DNA experiments be safer if they were done in a maximum-security lab, a P4 lab, in an isolated, nonpopulated area of the country? Question.

Would this be safer than using a P3 lab in one of the most densely populated cities in the nation? Question.

Is it true that in the history of science mistakes have been made, or [are] known to happen? Question.

Do scientists ever exercise poor judgment? Question.

Do they ever have accidents? Question.

Do you possess enough foresight and wisdom to decide which direction the future of mankind should take? Question.[15]

By this stage in the proceedings, the Harvard University delegation must have realized that they faced an uphill struggle, and in fact, a few minutes later a resolution was read out proposing a two-year ban on all rDNA

FIGURE 7.2
Alfred E. Vellucci—who served in Cambridge city government for forty years, including four terms as mayor—challenged the rights of MIT and Harvard to conduct research with potentially dangerous biological materials without public consent and oversight.
Source: Courtesy of the *Cambridge Chronicle*, file photo (October 23, 2002).

experimentation in Cambridge. This caused something close to an uproar. "If you pass that resolution," exclaimed Harvard biologist Mark Ptashne, "virtually every experiment done by members of the Biochemistry Department at Harvard will have to stop, and virtually every experiment done by about half the members of the Biology Department would have to stop, including experiments that no one, sir, *no one* has ever claimed had the slightest danger in them."[16] The resolution was voted down, but a number of scientists (including Harvard's Ruth Hubbard and MIT's Jonathan King) spoke against Harvard's proposal to build a P3 containment facility for rDNA research. It was left to MIT Dean of Science Robert Alberty to try to calm troubled waters by describing MIT's Center for Cancer Research. This facility, he told the meeting, had been designed to handle potentially hazardous organisms. But most important, he continued as follows:

No laboratory in the MIT Center for Cancer Research has ever been used for any recombinant work. People at the Center who would perform recombinant DNA experiments are among those who over the past two years in diverse public forums have urged the issuance of strict NIH guidelines to control various classes of recombinant DNA experiments and who in the interim have imposed upon themselves a voluntary moratorium on such experiments pending issuance of such guidelines.

Alberty assured the meeting that no rDNA work would be done at MIT until the Institute's Committee on the Assessment of Biohazards had studied the newly issued NIH guidelines. He concluded by extending an olive branch to the city by proposing a "joint mechanism with research institutions within the city for continuing assessment" of rDNA research.[17]

This first public hearing in Cambridge ended inconclusively, and a second one was quickly arranged for the following month. At this latter meeting, a resolution was passed establishing a special body—the Cambridge Experimentation Review Board (CERB)—to consider the question of whether rDNA experiments should be permitted in the city. A three-month moratorium on P3 and P4 experiments was imposed to give CERB time to do its job. In August the membership of CERB was announced, and it became obvious that the onus was on the research communities at MIT and Harvard to do exactly what Mayor Vellucci had suggested: explain themselves in the plainest of terms to a group of mostly lay citizens who had the power to recommend the prohibition of an entire field of biological research within the city limits.[18] Moreover, it was apparent that MIT rather than Harvard had the most to lose in this process. Looking back at the Cambridge hearings soon afterward, Baltimore recollected that

the problem was becoming partly an MIT problem, or had become partly an MIT problem. Actually, it became very clear early on that it was totally an MIT problem, because Harvard didn't have a facility in which they could do anything, and that we were the ones who were going to suffer from a moratorium.[19]

Throughout the second half of 1976, people around the world watched as the scientists and lay citizens of Cambridge publicly negotiated what some referred to as a "social contract" on the matter of rDNA research. CERB met on Tuesday and Thursday evenings throughout fall 1976, and the Tuesday meetings were open to the public. Many other events were organized by various interest groups. For example, "Science for the People" organized a teach-in at MIT on September 22; a week later, Harvard's George Wald and Matthew Meselson debated the issues as part of a regular public program called the Cambridge Forum. CERB members worked hard to inform themselves about the issues, attending frequent briefings and visiting relevant laboratories, including the P3 containment facility that had been created on the fifth floor of the Center for Cancer Research in the months following the Asilomar conference.[20] At the same time, MIT faculty and administrators worked hard to explain to the wider community the "who, what, why, where, and when" of rDNA research. As Phillip Sharp put it, "We made every effort we could to demystify this whole thing."[21]

A turning point in the controversy seems to have come toward the end of November 1976, when CERB organized a major debate between proponents and opponents of rDNA research. The aim was to create a format in which there would be time to dig beneath the surface of the various issues, and explore the risks and benefits of different kinds of research. CERB had this to report about the event:

In a five-hour marathon session, CERB carried out a type of mock courtroom affair. Board members served as a kind of jury while advocates on both sides of the issue presented their case, were given an opportunity to examine one another, and responded to questions raised by the "citizen jury." This format enabled the board members to evaluate how well scientists on each side of the controversy responded to the critical issues.[22]

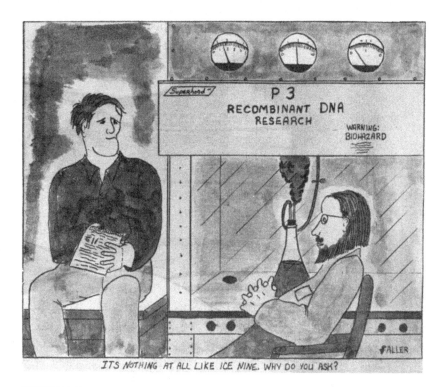

FIGURE 7.3

MIT's student newspaper, *The Tech*, ran a cartoon on February 15, 1977, likening research on rDNA to "Ice Nine," the fanciful substance in Kurt Vonnegut's 1963 science-fictional novel *Cat's Cradle*. In the novel a scientist creates Ice Nine, a unique form of ice that stays solid at room temperatures. After a tiny bit of the substance escapes from the laboratory, it ultimately kills all life on the planet. *Source*: Courtesy of *The Tech*.

A few weeks after this event, CERB released its findings at a public meeting. It essentially called for a regulation permitting rDNA research in Cambridge under the terms of the NIH Biosafety Guidelines together with a number of additional safeguards, including the requirements to apply to a local five-member Cambridge Biohazards Committee, and hold a public hearing and regular site inspections by local public health officials. After yet another public hearing, the Ordinance for the Use of Recombinant DNA Molecule Technology in the City of Cambridge was passed unanimously by the full city council on February 5, 1977, a little more than two and a half years after the publication of the Berg letter. Mayor Vellucci was delighted: "You can see, through all of our proceedings in the city government and city council," he crowed in an interview in May 1977, "that the little people did triumph over the big scientists or the big universities of the big United States of America."[23] For his part, though, Baltimore was struck by the fact that after a long, grueling, and highly unpredictable process, a group of "absolutely green citizens coming at this thing . . . end up in a position not far off from where the scientists ended up," adding, "I was surprised."[24]

THE MAKING OF "GENE TOWN"

In 1974, MIT found itself in a key position in relation to an emerging science, an emerging technology, and an emerging politics of science and technology policymaking. The decision to create the Center for Cancer Research had been pivotal, and from the beginning the center made big waves. In summer 1974, *Time* magazine included Baltimore in its list of "200 Faces from the Future," and the following year, he shared the Nobel Prize in Physiology or Medicine with Renato Dulbecco and Howard Temin "for their discoveries concerning the interaction between tumor viruses and the genetic material of the cell."[25] But not all of the publicity surrounding the Center for Cancer Research was quite so obviously positive. In particular, the public hearings in Cambridge City Hall had had the potential to result in the banning of rDNA research altogether, and if this had happened, it would certainly have relegated the new Center on Ames Street to minor league status. The temptation on the part of MIT scientists to lose patience with the long-winded public consultation that began in summer 1976 must have been great, especially for younger scientists like Sharp whose promising research careers hung in the balance. As it was, however, their patience and willingness to argue their case in public paid off handsomely.

Once the Cambridge ordinance was passed in 1977, the Institute found itself ideally placed to make rapid progress with rDNA-dependent research. Later in 1977, Sharp and his coworkers announced the discovery that the genes in advanced organisms are not continuous strings but instead contain "introns" that are deleted through a process of RNA splicing before messages are finally translated. (This is the work for which Sharp was to share the Nobel Prize in Physiology or Medicine in 1993.) By this time, too,

Robert Weinberg's group was hot on the heels of cancer-causing genes (oncogenes) in humans, and Baltimore's group was pioneering new techniques for cloning entire viral genomes in mammalian cells. What some students would later consider a "golden age" of research at MIT's Center for Cancer Research was then in full swing.

The innovations didn't stop there. In 1979, Baltimore started the negotiations that led to the creation of the Whitehead Institute at MIT. The biomedical entrepreneur and philanthropist Jack Whitehead took some years to decide on the right location for his new biological research institute, and his eventual offer to bring it to MIT was controversial within the faculty because he proposed an unusual model in which his Institute would be an independent but affiliated institution. In the end, this model was accepted by the MIT faculty. The Whitehead Institute opened in Kendall Square in 1982, and it quickly proved to be a huge success. The noteworthy point here is that the Whitehead Institute would never have gotten off the ground if the early success of the Center for Cancer Research had not greatly reinforced MIT's strengths in molecular genetics and rDNA research.[26]

In this way, the events of the late 1970s led to the further and rapid growth of the life sciences at MIT: at a stroke, the creation of the Whitehead Institute expanded by almost a third the number of faculty positions in the Department of Biology. This, in turn, led to MIT's growing involvement in human genomics under the leadership of Whitehead fellow Eric Lander. The Whitehead Institute/MIT Center for Human Genomics was created in 1990, and it came to play a pivotal role in the Human Genome Project, eventually creating the conditions for the establishment in 2004 of a new and separate genomic research facility, the Broad Institute of MIT and Harvard. By this time, MIT had been a world-leading center for genetic and genomic research for almost a generation.

If the payoff for science from the rDNA controversy was considerable, so, too, was the payoff for the emerging industrial sector of biotechnology. From early in the rDNA debate, it was realized that the newfound ability to manipulate genes in microorganisms had enormous commercial potential. As early as 1976, biologist Herbert Boyer (University of California at San Francisco) joined with MIT alumnus and venture capitalist Robert A. Swanson to create Genentech, a company devoted to the exploitation of rDNA technology in the commercial manufacture of medically useful

FIGURE 7.4
David Baltimore (left), Gerald Fink (middle), and Jack Whitehead visit the construction site of MIT's Whitehead Institute for Biomedical Research, ca. 1980. Baltimore helped convince Whitehead to donate the funds to MIT in 1979, and the center opened in 1982. Baltimore served as its first director, from 1982 to 1990; Fink later served as director from 1990 to 2001.
Source: Courtesy of the Whitehead Institute.

molecules such as insulin and growth hormone. At this early stage, it was already clear that the San Francisco Bay Area would be one major locus for the growth of the new biotechnology, but would the Bay State—and Cambridge, in particular—be another? Once again, if the city of Cambridge had turned its back on rDNA technology, the answer would almost certainly have been no; but the passage of the rDNA ordinance in 1977 actually facilitated the emergence of a second major biotechnology cluster, centered in Kendall Square, right alongside the burgeoning life sciences laboratories of MIT.

Because Cambridge—unlike Boston as well as the other surrounding towns and cities—had its local regulatory framework in place, companies thinking of setting up in the region knew what to expect and what was expected of them. Moreover, civic leaders in Cambridge were already familiar with many of the ideas, issues, and on occasion, even the individuals involved in proposals to start up biotechnology companies. The classic case here is Biogen, the company that was launched in 1978 by, among others, Harvard University's Walter Gilbert and MIT's Phillip Sharp. In 1980, Biogen chose Cambridge as the ideal location for its commercial and managerial operations, citing not only the proximity to Harvard and MIT but also "the fact that the city, as a result of the initial work of the [Cambridge Experimentation Review] Board, has made the political and scientific decision to permit the use of recombinant DNA techniques within the framework of the City's Ordinance and the NIH Guidelines."[27] Two years later, at the ribbon-cutting ceremony to celebrate the opening of Biogen's research division in Cambridge, none other than Mayor Vellucci announced that he had "no fear of recombinant DNA as long as it paid taxes."[28]

The Biogen story was repeated many times over in the 1980s and 1990s as companies with names like Genetics Institute, Genzyme, Millennium, and Vertex chose to base their operations in the small city that was obviously emerging as one of the world centers of biotechnology. As early as 1985, the Massachusetts Biotechnology Council (MBC) was created as the first biotechnology trade association in the United States. Today the MBC represents the interests of more than 550 biotechnology companies, universities, and academic institutions.[29] Although the member organizations are located throughout the Greater Boston metropolitan area, they are particularly concentrated in the Kendall Square area. What MIT President Howard Johnson (1966–1971) described as "a desolate place with little human presence after five p.m." is now the epicenter of what the MBC describes as "a world premier life sciences supercluster."[30] Of course, many factors—including the academic strengths of the city of Cambridge, the financial strengths of neighboring Boston, and the affordability of property in the Kendall area—contributed to the creation of "gene town," as it is known to many. But one crucial factor was certainly that as early as the late 1970s, Cambridge knew where it stood on the key issues at the dawn of the age of modern biotechnology.

AND THE MORAL OF THE STORY IS . . .

There is no single or simple lesson to be learned from the rDNA controversy in Cambridge. Some of the scientists who led the call for the initial moratorium on rDNA research never doubted that they had been right to do so, even if (as in the case of Baltimore) they were surprised by the subsequent course of events. Others (including Watson), however, soon repented of their actions when they saw what looked to them like ill-conceived and potentially disastrous attempts by local communities to seize control over the conduct of research from the academic and scientific communities. Conversely, some observers of CERB saw its activity as an almost cynical exercise in the management (read containment) of popular dissent, while others viewed it as an inspirational model of how citizens can engage constructively with scientists in decision making on socially and politically sensitive areas of research. Interestingly, since the mid-1970s there has been a great deal of experimentation with different models of citizen participation in science and technology assessment, but much of this has taken place outside the United States—for example, in Australasia, Canada, and Western Europe.[31]

For the present purposes, I prefer to draw a moral from this story that is at once more parochial and more personal. At a critical moment in its history, MIT faced the need to explain itself to the wider community in order to secure a license to continue operating in a promising area of research and innovation. The Institute rose to this challenge. It did not retreat into the ivory tower. Instead, it reached out in a process of constructive engagement with citizens and community representatives—a process that operated under Mayor Vellucci's unforgettable injunction to "refrain from using the alphabet." The process was far from perfect, and the outcome was, for a time, extremely uncertain. In the end, though, the Institute secured its license to operate and was able to move forward in ways that were scarcely imaginable just a few years earlier. For me, the moral of this story is obvious: MIT is all the stronger for having close and constructive relationships with the wider community. Fortunately enough, it pays to be a socially responsible citizen.

ACKNOWLEDGMENTS

At an early stage in this project, Victor McElheny provided an extremely useful orientation on the history of the life sciences at MIT. In addition, the author gratefully acknowledges the following colleagues who generously agreed to be interviewed: David Baltimore, Louis Chodosh, Herman Eisen, Nancy Hopkins, David Housman, Margarita Siafaca, Phillip Sharp,

Cliff Tabin, Charles Weiner, and Robert Weinberg. Michael Rossi contributed important research assistance.

NOTES

1. Sheldon Krimsky, *Genetic Alchemy: The Social History of the Recombinant DNA Controversy* (Cambridge,MA: MIT Press, 1982), 82. For other secondary sources that have been used in constructing this account, see also Charles Weiner, "Drawing the Line in Genetic Engineering: Self-regulation and Public Participation," *Perspectives in Biology and Medicine* 44 (Spring 2001): 208–220; Charles Weiner, "Recombinant DNA Policy: Asilomar Conference," in *Encyclopedia of Ethical, Legal, and Policy Issues in Biotechnology,* ed. Thomas J. Murray and Maxwell J. Mehlman (New York: Wiley, 2000), 2:210; Susan Wright, *Molecular Politics: Developing American and British Regulatory Policy for Genetic Engineering, 1972–1982* (Chicago: University of Chicago Press, 1994).

2. Maxine Singer and Dieter Soll, "Guidelines for DNA Hybrid Molecules," *Science* 181 (September 21, 1973): 1114.

3. David Baltimore, "Where Does Molecular Biology Become More of a Hazard Than a Promise?" lecture to the Technology Studies Workshop at MIT, November 6, 1974, MC100, box 38, folder 575:6, Recombinant DNA History Collection, MIT Institute Archives and Special Collections, Cambridge, Massachusetts.

4. Paul Berg et al., "Potential Biohazards of Recombinant DNA Molecules," *Science* 185 (July 26, 1974): 303.

5. Key names here include Jonathan Beckwith (Harvard), Jonathan King (MIT), Ruth Hubbard (Harvard), and George Wald (Harvard).

6. Interview with David Baltimore by Charles Weiner and Rae Goodell, May 13 and July 22, 1975. Transcript available in MC 100, Box 1, Folder 5:37–38, Recombinant DNA History Collection, MIT Institute and Special Collections, Cambridge, Massachusetts.

7. Author's interview with Phillip Sharp, January 8, 2009.

8. For detailed treatments of these developments, see Krimsky, *Genetic Alchemy*; Wright, *Molecular Politics*.

9. Press release, June 23, 1976, U.S. Department of Health, Education, and Welfare. Copy in the Donald S. Frederickson Papers, National Library of Medicine,

Bethesda, MD, available at <http://profiles.nlm.nih.gov/FF/Views/Exhibit/documents/rdna.html> (accessed June 9, 2009).

10. Alfred Vellucci, in "Hearing on Recombinant DNA Experimentation, City of Cambridge," June 23, 1976, MC100, box 33, folder 458:3, Recombinant DNA History Collection, MIT Institute Archives and Special Collections, Cambridge, Massachusetts.

11. "A Scientific Breakthrough," *Washington Post*, July 2, 1976; available at <http://profiles.nlm.nih.gov/FF/Views/Exhibit/documents/rdna.html> (accessed June 9, 2009).

12. Coverage of the meeting at Harvard appeared in a local newspaper, the *Boston Phoenix*, on June 8, 1976. For relevant details, see Krimsky, *Genetic Alchemy*, 299–300; Maryann Feldman and Nichola Lowe, "Consensus from Controversy: Cambridge's Biosafety Ordinance and the Anchoring of the Biotech Industry," *European Planning Studies* 16, no. 3 (April 2008): 395–410.

13. Alfred Vellucci, quoted in John Kifner, "'Creation of Life' Experiment at Harvard Stirs Heated Dispute," *New York Times,* June 17, 1976.

14. Alfred Vellucci, in "Hearing on Recombinant DNA," 3–4.

15. Alfred Vellucci, in ibid., 37–39.

16. Mark Ptashne, in ibid., 73.

17. Robert Alberty, in ibid., 120–122.

18. Cambridge was not the only city to debate rDNA policy in the late 1970s. In total, between 1975 and 1979, nine cities and towns across the United States considered proposals for the local regulation of rDNA research, including Ann Arbor, Michigan, Princeton, New Jersey, and San Diego, California. See Krimsky, *Genetic Alchemy*, 294–391.

19. David Baltimore, interview, May 3, 1977. Transcript available in MC100, box 1, folder 6:49, Recombinant DNA History Collection, MIT Institute Archives and Special Collections, Cambridge, Massachusetts.

20. Author's interview with Phillip Sharp.

21. Ibid.

22. Feldman and Lowe, "Consensus from Controversy," 401.

23. Alfred Vellucci, interview, May 9, 1977. Transcript available in MC100, box 14, folder 163:3–4, Recombinant DNA History Collection, MIT Institute Archives and Special Collections, Cambridge, Massachusetts.

24. Baltimore, interview, May 3, 1977, 60.

25. For information on their Nobel Prize, see <http://nobelprize.org/nobel_prizes/medicine/laureates/1975/index.html> (accessed June 9, 2009).

26. Baltimore became the first director of the Whitehead Institute in 1982; several other Center for Cancer Research faculty, including Weinberg, joined him at Whitehead Institute that year.

27. Letter to Robert Neer, chair, Cambridge Experimentation Review Board, December 30, 1980, in Feldman and Lowe, "Consensus from Controversy," 403.

28. Alfred Vellucci, quoted in Feldman and Lowe, "Consensus from Controversy," 405.

29. For the Massachusetts Biotechnology Council, see <http://www.massbio.org> (accessed June 9, 2009).

30. Howard Johnson, *Holding the Center: Memoirs of a Life in Higher Education* (Cambridge, MA: MIT Press, 1999), 139; Massachusetts Biotechnology Council, <http://www.massbio.org> (accessed June 9, 2009).

31. For more information, see Edna Einsiedel and Deborah L. Eastlick, "Consensus Conferences as Deliberative Democracy: A Communications Perspective," *Science Communication* 21 (2001): 323–343; Edna Einsiedel, E. Jelsøe, and T. Breck, "Publics at the Technology Table: The Consensus Conference in Denmark, Canada, and Australia," *Public Understanding of Science* 10 (2001): 83–98; articles in S. Joss, ed., *Public Participation in Science and Technology*, special issue of *Science and Public Policy* 26 (1999): 290–373.

LOTTE BAILYN

On Sunday, March 21, 1999, the front page of the *Boston Globe* carried an article with the following headline: "MIT Women Win a Fight against Bias. In a Rare Move, School Admits Discrimination."[1] The newspaper story followed a frantic week of behind-the-scenes effort at MIT to get a report to faculty members before they read about it in the newspaper. The report, "A Study on the Status of Women Faculty in Science at MIT"— now everywhere referred to as "the MIT Report"—was emailed to the faculty on Friday, March 19, with the warning that the *Globe* was coming out with the story.[2] The following Tuesday, the *New York Times* carried the story with its own front-page headline: "MIT Admits Discrimination against Female Professors."[3] By then it was spring vacation at MIT, and things were pretty quiet for everyone except the protagonists; they were deluged with emails, phone calls, and requests to speak from all over the country—indeed, from all over the world.

What the MIT Report showed, and what was picked up by these newspaper stories, was that even highly successful women scientists, members of the National Academies and widely known for their research, were subject to subtle unintentional discrimination. It was twenty-first-century discrimination—not blatant harassment—which according to the report, "consists of a pattern of powerful but unrecognized assumptions and attitudes that work systematically against women faculty even in the light of obvious good will."[4] The conclusion was based on extensive interviews with the tenured women professors in the School of Science that revealed their exclusion from large multi-investigator research projects, absence from important committees, difficulties in getting the space and other resources they needed for their research, and inability to control their teaching assignments—all standard perquisites of senior faculty members.[5] Because of this subtle bias, their salaries were lower, their space was less, and they had to work harder for the resources they did get. The MIT Report also showed that in the previous twenty years, the percentage of women faculty in MIT's School of Science had not significantly increased and there had never been a woman in any position of leadership in that school, not even as associate department head or associate center director.

It is hard to describe the intensity of the response that followed the public dissemination of the MIT Report. Messages came from everywhere, but mainly from women and institutions anxious to hear more, and asking for help to do similar studies. Others just wanted to thank MIT for finally acknowledging what they had always known in some way yet had not been able to express easily. One ten-year-old girl wrote in appreciation, "You are opening the door of opportunity even wider for girls like me who want to go as far as possible in science."[6] Some women reported on administrations that finally looked at their salary data and realized that, yes, there was a gender gap—something that was known statistically but had not previously moved anyone to action. A long article in *Science* reported on the MIT story as well as the experience of women scientists at Harvard.[7] Professor Nancy Hopkins, who initiated the study that led to the report, and dean Robert Birgeneau (now chancellor of Berkeley) were invited to the White House on April 7, 1999 (Pay Equity Day); President Bill Clinton and Mrs. Clinton congratulated MIT for identifying an important problem.[8] Hopkins accepted sixteen other invitations to speak that year, and in December 1999, when the *Chronicle of Higher Education* reported in detail on how this all came about, the invitations multiplied.[9] These invitations and inquiries to her and others involved in the report continue to this day. For example, a letter from an eighth grader in New York City asking for more information arrived in 2008—almost ten years after the news broke.[10]

Support also came from funding agencies. The Ford Foundation offered MIT a grant if the university would pull together other institutions to spread the word. In the end, Ford and Atlantic Philanthropies gave a total of one million dollars to the effort. To meet the goals of these grants, MIT President Charles Vest (now President of the National Academy of Engineering) launched several initiatives, including inviting eight university presidents (from Berkeley, Caltech, Harvard, Michigan, Penn, Princeton, Stanford, and Yale) to a conference in early 2001.[11] Three of these institu-

FIGURE 8.1

After MIT released its report, "A Study on the Status of Women Faculty in Science at MIT" in 1999, the *Chronicle of Higher Education* ran a lengthy article on the issue, featuring this photograph on its front page. Standing, left to right, are professors Sylvia Ceyer, Paola Rizzoli, Penny Chisholm, Nancy Hopkins, Leigh Royden, JoAnne Stubbe, and Mary-Lou Pardue.

Source: Courtesy of Rick Friedman Photography.

tions had initially commented on the media reports by commending MIT on its response to this unfortunate situation but asserting that it did not apply to them. In the end, though, they all acknowledged the problem and vowed to work on it.

There is only one way to explain this extraordinary response: this was an issue waiting to be heard. It had appeared in dozens of reports sitting on the shelves of presidents, provosts, deans, and department heads in universities around the country—reports that resulted in no action. Yet when women scientists who had been successful at the country's leading science and technology university described their experiences, the issue became a legitimate reality not only for women in universities but for all professional women. And when MIT's President stated in his introductory comments to the MIT Report that "I have always believed that contemporary gender discrimination within universities is part reality and part perception. True, but I now understand that reality is by far the greater part of the balance"—a statement picked up in all the newspaper accounts—the genie was out of the bottle. As noted in an editorial in the *San Francisco Chronicle* on March 24, 1999,

> But for one significant difference, a Massachusetts Institute of Technology study confirming discrimination against women faculty members would probably have been ignored by college administrators across the country—like so many similar reports.
>
> The difference this time, however, is that the respected president of MIT— one of the most prestigious universities in the nation—not only did not ignore the report, he acknowledged existence of the discrimination and took steps to redress it.[12]

On March 28, the *New York Times* ran an editorial asserting that "the MIT study is unusual because it examines tenured women who have excelled in the male-dominated sciences, and whose collective experiences with bias cannot be explained away by special circumstances." It commended MIT for having "confronted this reality boldly and . . . taking steps to correct the inequities and improve hiring practices."[13]

No public event like this could pass without its detractors. A first admonition came in April from the legal editor of the *San Francisco Chronicle* who warned that "in today's legal system, those who are candid enough to admit their own mistakes may find the information used against them."[14] MIT had no general counsel at that time and had not asked for legal advice before allowing the report to become public. The feeling was that admitting mistakes and rectifying them would actually lessen the probability that anyone would take legal action against the Institute.

More was to come. On December 14, 1999, a professor of psychology at the University of Alaska in Fairbanks released a report titled *MIT Tarnishes Its Reputation with Gender Junk Science.*[15] It was distributed by the Independent Women's Forum, a conservative research and educational institution that described its mission this way:

> To rebuild civil society by advancing economic liberty, personal responsibility, and political freedom. IWF builds support for a greater respect for limited government, equality under the law, property rights, free markets, strong families, and a powerful and effective national defense and foreign policy.[16]

A *Wall Street Journal* editorial on December 29, 1999, built on the professor's document to denounce the MIT Report and the studies of universities that were following its guidelines as "politicized exercises in 'social science.'"[17]

The problem with the MIT Report, according to these critics, was primarily that there was an alternative explanation for the dearth of women faculty in the sciences: namely, women are less interested in entering science and may have different aptitudes that lead more easily into other fields more congruent with their values. Further, there was a sense that the women scientist complainants were both judge and jury, and "opted for a university-funded study rather than taking their complaints to court."[18] Finally, the report was deemed "junk science" because it did not contain the details of the data that led to the conclusion of subtle discrimination.

The alternative explanation—that women do not enter science because they lack interest or aptitude—is not relevant to the present case. The women involved in the MIT Report did have a deep commitment to science and did have the ability for it; they had the motivation and perseverance to pursue it as a career, and were successful in that pursuit. It was their experiences—out of line with their accomplishments—that led to the report's conclusion. Moreover, the internal study was specifically undertaken in order *not* to create an adversarial legal confrontation around these concerns.

But what about the data? It is true, the report did not contain any of the data that were collected; nor were these data released by MIT. The most telling information was the individual experiences of the women, which had been gathered under a promise of anonymity, and with only seventeen women spread across five departments, these women would have been readily identifiable. So the report that was released was a narrative account not of data but rather of a process. And it is this process that has been copied by universities around the country and the world: work collectively to look carefully at your own data, and work collaboratively with the administration to understand the data and resolve inequities that have been uncovered. To this day, requests continue to come to MIT for help in setting up equivalent processes at other institutions.

Reports from the other four schools at MIT were released in 2002 and confirmed the findings in the School of Science.[19] For example, the report from the Sloan School of Management had a careful analysis of the senior women compared to a matched group of senior male faculty that confirmed the less satisfactory experience of the women faculty members compared to their male pairs.[20] Ironically, though these reports did present full data and full analyses, they received no outside acclaim.

And so the outside world reacted to the MIT Report. But how did MIT respond? When faculty returned after spring vacation in 1999, the women members had a celebratory get-together. The male faculty made hardly any response. The administration, on the other hand, put into place a series of changes that endured and continue to evolve to this day. Gender had been put on the table, and the administration began to respond.

WORDS INTO ACTION

Almost immediately, MIT Provost Robert Brown (now President of Boston University) asked the deans in each of the other MIT schools to put together a committee to do a study of their schools, following a similar process as the School of Science.[21] Each school went through the process of interviewing the women faculty members, and collecting data on salary, space (where applicable), teaching assignments, committee participation, awards and honors, and so on. As already indicated, these reports were released in 2002 and are available online.[22] Although the methods used differed somewhat from school to school (as did the form of the output), the reports detailed the same kind of inequities and marginalization that had been found in the School of Science.

The chairs of these "gender equity committees," along with some of the School of Science faculty members involved in the MIT Report, met monthly with Nancy Hopkins starting in fall 1999 to keep the momentum going, and share experiences from their schools and consider issues that need attention. They have continued to meet as a group, now with the new coassociate provost for faculty equity. Also, each chair works in her school to monitor faculty salaries annually, and keep an eye on equity issues for women and men as well as underrepresented minorities.

The Institute's administration was equally concerned to make progress—indeed, the publicity had in effect given MIT the responsibility to move on this issue. In 2000, the President and Provost created the Council on Faculty Diversity. Its goal was to "work with the faculty, departments, schools, and the senior administration to help the Institute aggressively promote faculty diversity. These efforts will work to establish a *sustained* institutional environment that will attract a diverse faculty that reflects the students we educate."[23]

The new council was to be cochaired by Provost Brown, Hopkins, and Associate Provost Phillip Clay.[24] Both Brown and Clay were members of the Academic Council, the chief policymaking group of the Institute. A faculty subcommittee of the Academic Council goes over every promotion and tenure case at MIT, and also monitors salaries, so membership in the Academic Council is a position of some influence. When the Council on Faculty Diversity was established, Hopkins was made a full member of both the Academic Council and this subcommittee—an unusual step, signaling the administration's commitment to faculty diversity.[25] All this happened as of July 1, 2000.

The members of the Council on Faculty Diversity consisted of the chairs of the gender equity committees, faculty members from underrepresented minority groups, and administrators. One subcommittee, chaired by the head of the engineering gender equity effort, worked on hiring and developed guidelines to ensure more diversity in the hiring process of junior faculty.[26] These guidelines were distributed by the provost to all department heads, who were asked to provide them to the search committees in their departments. Another subcommittee worked on a new set of family policies, including automatically extending the tenure clock for women who bear a child, providing for parental leave for a new child in the family, and permitting part-time arrangements for tenured faculty who need time for the care of a child, parent, or partner.[27] These policies were approved by the Academic Council in December 2001, and were amended in 2006 to include the possibility for men and adoptive parents to request a tenure extension as well as for women with a second child to request a second extension.[28]

At the same time, President Vest started enlisting his peers in the eight other institutions already mentioned for a conference called Gender Equity in Academic Science and Engineering. It was a way of disseminating the concerns and sharing efforts at solutions. That first conference was held at MIT on January 28–29, 2001. It brought together presidents, chancellors, provosts, and women faculty members in science and engineering from the nine universities. It ended with a "pledge of commitment" indicating the presidents' recognition that "barriers still exist to the full participation of women in science and engineering." They agreed that their institutions would work toward three goals: faculties should reflect the diversity of the students they educate; women faculty members should have equity and full participation; and no faculty member—male or female—should be disadvantaged by family responsibilities. The presidents agreed to analyze salaries and other resources provided to women faculty members at their universities, and reconvene "to share the specific initiatives we have undertaken to achieve these objectives." They also recognized "that this challenge will require significant review of, and potentially significant change in, the procedures within each university, and with the scientific and engineering establishments as a whole."[29]

Prior to the presidents coming together again, the women faculty members from these universities collected data and met to share this infor-

FIGURE 8.2
Discussions at the first conference on Gender Equity in Academic Science and Engineering, hosted by MIT in January 2001. At this table, the conversation included Lee Bollinger (left), President of the University of Michigan; Richard Levin, President of Yale University; Jan de Vries, Vice Provost of the University of California at Berkeley; Professor Nancy Hopkins of MIT; Professor Shirley Tilghman of Princeton University; and Thomas L. Magnanti, MIT Dean of Engineering. Five months after this photo was taken, Tilghman became the first female President of Princeton.
Source: Donna Coveney, courtesy of the MIT News Office.

mation, which they were then able to bring back to their administrations. A second presidents' conference was held three years later (in Washington, DC, on April 17–18, 2004) in conjunction with the American Association of Universities meetings. It was hosted by President Vest along with the presidents of Princeton and the University of Michigan. In subsequent years the women faculty members again got together. And it was at the conference in Ann Arbor, Michigan, in 2006 that the non-MIT participants decided to call this group of universities "the MIT9." True to pattern, there was another presidents' conference in 2007, again in conjunction with the American Association of Universities meetings. This time Harvard was the host, and the topic was faculty of color.

There were also ramifications of the MIT Report in a wider sense. The National Science Foundation rethought its approach to helping women advance in science. Previously it had emphasized individual fellowships, but recognizing these more systemic issues, in 2001 it started the ADVANCE program—a program of significant grants to universities for institutional transformation. In 2005, the National Academies charged a "committee on maximizing the potential of women in academic science and engineering," chaired by Donna Shalala (President of the University of Miami, and former secretary of the U.S. Department of Health and Human Services), to pull together all the relevant research and programs that could be helpful in this endeavor.[30]

Finally, what effect was there on the number of women faculty members in the schools of Science and Engineering? An analysis in the March–April 2006 issue of the *MIT Faculty Newsletter* showed that the number of women in the School of Science rose in the early 1970s from almost none to near 20 (compared to some 260 men) in response to equal opportunity legislation.[31] It then stayed at that level for almost twenty years. After the original confidential report on women faculty in the School of Science was given to Dean Birgeneau in 1996, the number rose to about 35 by 2000 (compared to roughly 230 men). That figure then served as the plateau for a few years, even as the number of men increased to 240. In 2008, however, the numbers were 45 women and 236 men. In other words, since the original data were presented to the Dean of the School of Science, the number of women has more than doubled.

In the School of Engineering, there was a fairly steady increase in the number of women faculty from under 20 in 1990 to 30 in 2000 (compared to 311 men). When this school's report was presented to the dean in 2001, the number rose more steeply until 2005, when there were about 50 women and 312 men. Since then the number of women has stayed fairly constant, with 52 women and 320 men in 2008. Furthermore, whereas there had been only one woman in an administrative capacity in the schools of Science and Engineering before the MIT Report was made public, within two years, the number of women in top decision-making administrative posts was more than 10 and soon would include the first woman President of MIT, Susan Hockfield.[32]

So there has, indeed, been progress. It is also clear that the repercussions of the MIT Report continue to this day, both at MIT and elsewhere. So how did it all happen? What led to that original report in 1996, and how was it transformed into the report that was picked up by the newspapers in 1999?

HOW IT HAPPENED—AND WHY

It started in 1994.[33] Nancy Hopkins, a professor of molecular biology at MIT for twenty years, had been trying for some time to get more lab space for a new avenue of research that she was undertaking. She had little luck with the director of her center. She grew frustrated and wondered why it was so difficult for her to pursue this new research; others around her seemed to be having an easier time. On top of that, a course she had developed was given to someone else to teach. Hopkins had never been a feminist; she was concerned only with being a scientist, having even decided not to get married or have children in order to pursue that goal. Yet slowly, during those frustrating years, she began to think that perhaps this was not a fault of her own but instead resided in bias against her. She collected all the evidence she had about her attempts to get the resources she needed and consulted a lawyer. He confirmed that there was a case for harassment as defined by MIT, and probably for discrimination, and urged her to contact the higher administration.

Hopkins drafted a letter to President Vest, asserting that there was discrimination afoot and providing some of the evidence. But she was concerned about whether the President would take the letter seriously or be offended. She decided, with some hesitation, to check with a colleague, Professor Mary-Lou Pardue, a respected senior biologist who had been a

FIGURE 8.3

MIT professor Nancy Hopkins, shown here with some of her laboratory specimens in 1999. Hopkins uses zebrafish as a model organism for studying the genetics of cancer.

Source: Courtesy of Rick Friedman Photography.

member of the National Academy of Sciences since 1983. They had never talked about these things, and Hopkins feared that Pardue would think ill of her. When, in contrast, Pardue agreed with the contents and asked to sign the letter as well, everything changed.

Together they decided to check with all the other senior women in the School of Science and did so, again with hesitation. Sixteen of the seventeen women they approached agreed to sign the letter.[34] At this point they changed tactics again, and decided to send their letter to the new Dean of Science, Robert Birgeneau, and ask him to set them up as a committee to investigate the data more thoroughly. They sought to document the discrimination they had seen and felt along with its professional consequences for them.

The dean, however, ran into "extreme skepticism" from his department heads and center directors when trying to create this committee, and was at first reluctant. He had been using a consensus-driven style, and now he discovered that some of these people were not at all in favor of such a committee. But with direct and explicit support from President Vest, he did establish a preliminary committee to construct a charge. In the end, with the go-ahead from this committee, he appointed six senior women to a new committee charged with collecting these data and augmented it with three senior male faculty members. Despite initial hesitation on the part of the women, this turned out to be useful since these men—all of whom were or had been department heads—knew how the system worked, which the women did not.

Looking back on this period, Birgeneau recalls a meeting in his dean's conference room with fifteen of these women—a meeting that had a great impact on him:

Listening to the personal stories of all 15, one at a time, was simply overwhelming. At that moment I realized that there really was a systemic problem and that it needed to be addressed immediately not just for the health and welfare of these women faculty but also for the health of MIT as a whole.[35]

Once the committee was appointed, they interviewed, with a promise of absolute confidentiality, all of the women faculty members in the School of Science.[36] Members as much as possible also collected data on salaries, space, and other resources. On the basis of this material, they compiled a detailed report, providing specific incidents in each of the six departments of the school. This report was given to the dean in 1996, and was the confidential report that was available only to the dean, provost, and president (although relevant selections went to the department heads as well). In the meantime, the dean had already begun to remedy some of the inequities found in salary and lab space.

Although the women were now in better shape and back at their research, there was a continuing feeling that the rest of the faculty should be informed, if only to ensure that progress would continue. The School of Science was clearly not unique, and the understanding gained by the

FIGURE 8.4

In 1991, physicist Robert Birgeneau became the new Dean of Science at MIT. When Professor Nancy Hopkins and some female colleagues in MIT's School of Science raised concerns about gender discrimination a few years later, Birgeneau convened a special committee to investigate, leading to the MIT Report in 1999.

Source: Donna Coveney, courtesy of the MIT News Office.

committee would be useful for the whole Institute. No one wanted to share the original, highly detailed report more widely. The Dean of Science would not allow it, and the women didn't want it. Many of the episodes portrayed were humiliating, and the women—and even some of the men involved—would have been embarrassed to have this information made public. In spring 1997, there was an informal discussion in the faculty policy committee of what had happened, but that was confidential and did not go further. So the feeling of wanting to share persisted. In 1998 Professor Mary Potter, the new cochair of the committee of women faculty in the School of Science, made a shortened version, aggregated across departments, with individual data eliminated, but it, too, was not deemed ready for the public.[37] In 1999, with a woman chair of the faculty, another discussion was held in the faculty policy committee with the explicit goal of finding a way to make some of this information available to all MIT faculty.[38] On the basis of this discussion, the new faculty chair asked Hopkins to write a different kind of report, a story of the process they had gone through and the response of the dean, including only the recommendations from the original draft. In addition, the faculty chair, Dean, and President agreed to write introductory comments. This version was acceptable to all, and it was decided to publish it as a special edition of the *Faculty Newsletter*.[39] But while that publication was being readied, the *Globe* began to call and precipitated the frantic week alluded to at the beginning of this chapter.

While giving a talk to the Knight Fellows (science journalists who come to MIT for a year of study), Hopkins was asked by Andrew Lawler from *Science* what it was like to be a woman scientist at MIT. Her response contained a mention of the coming report and a question about its possible newsworthiness. That led to a call to Boyce Rensberger, head of the Knight program, who alerted Carey Goldberg of the *New York Times* and Kate Zernike of the *Boston Globe*. Zernike responded first and got the original story.

That is the story of how the MIT Report came about. How can we understand the traction of this report compared to so many others that ended up on shelves? A previous analysis listed the factors that had to come together to create the results described.[40] These factors can be summarized as five critical conditions and events.

First, the presence of a champion, Hopkins, who persevered and managed the process at all points. Second, the point of connection with Pardue, and the way that Hopkins worked collectively with the women faculty members in the School of Science throughout the process. Third, Birgeneau's understanding and positive response along with Vest's continuous support. Fourth, finding a format for the original report that could be made public. And last but not least, the media, which clearly catalyzed the national and international response. It is interesting to speculate what the MIT response would have been to just the publication in the *Faculty Newsletter* without the front-page stories in the *Boston Globe* and *New York Times*. The administration was clearly committed, and individual inequities were being addressed. But would there have been systemic change? Most likely "it really took all of these factors to bring about the positive outcome that resulted from the MIT Report. It was a combination of [good]will, persistence, good timing, and good luck."[41]

NOW AND INTO THE FUTURE

Where are we now? Gender is certainly on the table, and the number of women is going up. Many of the recommendations from the original report have been put into practice. There are new family policies, and ways to subsidize faculty for child care needs are being considered. There is also some acceptance of the existence of evaluation bias; we are not yet a true meritocracy, and there is training being given to department heads and search committees on the dynamics of this bias. Women are in major decision-making administrative positions, and involved in hiring and promotions; committees are in place to monitor salaries and other equity issues. We are still getting requests from faculty members in other universities for help in their investigations, and are still hearing reports of administrations unwilling to share data.

In conservative circles, there continues to be an unwillingness to accept the existence of subtle gender bias or see it as having anything to do with the relatively low, although improving, number of women scientists on university faculties. The MIT Report continues to be seen by some as junk science that supports political correctness. Even data in a refereed article in *Nature* are being questioned.[42] Meanwhile, other research universities, with

FIGURE 8.5

MIT President Charles Vest supported the effort that went into the MIT Report in 1999, publicly endorsed its findings, and organized colleagues from peer institutions to continue to investigate gender and racial discrimination in higher education. Vest (center) and his wife, Rebecca (right), are shown here hosting President Bill Clinton (left) in spring 1998, when Clinton delivered the MIT commencement address.

Source: Donna Coveney, courtesy of the MIT News Office.

the help of the National Science Foundation's ADVANCE grants, are becoming perhaps more innovative (Berkeley and Michigan are prime examples) by trying new procedures such as part-time tenure tracks.

Both MIT and the MIT9, although continuing to work on gender, are now moving to engage the issue of faculty members from underrepresented minorities. Indeed, in June 2007 MIT embarked on the Initiative for Faculty Race and Diversity, which is gathering interview, survey, and cohort data on the experiences of underrepresented-minority faculty members at MIT in order to provide recommendations and an implementation plan for increasing minority recruitment and retention.[43] MIT also has two associate provosts for faculty equity—one white woman and one African American man.[44] In both cases, it is important for MIT to work collaboratively with other universities to increase the pool of applicants for faculty positions from all groups.

The effort to make university faculties fully inclusive is a continuous process that is necessary and important not only for the individuals involved but also for the universities as well as society at large. The MIT Report has been and continues to be a player in this evolution.

ACKNOWLEDGMENTS

I would like to thank Nancy Hopkins, Doreen Morris, Mary Potter, and Lydia Snover for providing me with material for this chapter, and the staff of the MIT libraries who helped track down newspaper reports. I would also like to thank Robert Birgeneau, Robert Brown, Nancy Hopkins, Mary-Lou Pardue, Mary Potter, and Charles Vest for helpful comments on an earlier draft.

NOTES

1. Kate Zernike, "MIT Women Win a Fight against Bias," *Boston Globe*, March 21, 1999.

2. "A Study on the Status of Women Faculty in Science at MIT," *MIT Faculty Newsletter* 11, no. 4 (March 1999); available at <http://web.mit.edu/faculty/reports/sos.pdf> (accessed December 24, 2008); hereafter referred to as the MIT Report.

3. Carey Goldberg, "MIT Admits Discrimination against Female Professors," *New York Times*, March 23, 1999.

4. MIT Report, on 12. See also Virginia Valian, *Why So Slow? The Advancement of Women* (Cambridge, MA: MIT Press, 1998), for an extensive review of the research underlying the subtle discrimination that resides in gender schemas held by both women and men; and Susan Sturm, "Second Generation Employment Discrimination: A Structural Approach," *Columbia Law Review* 101 (2001): 458–568, which differentiates, from a regulatory point of view, second generation discrimination ("a subtle and complex form of bias," 458) from first order discrimination ("overt exclusion, segregation of job opportunity, and conscious stereotyping," 465).

5. These interviews were conducted by a committee convened by the Dean of Science that consisted of six senior women (one from each department in the

School of Science that had one, plus one other) and three men. The formation of this committee is explained in a later section of this chapter. The committee also collected data on salaries and grant support, space, committee membership, and other aspects of university life.

6. Letter to Nancy Hopkins, December 12, 1999.

7. Andrew Lawler, "Tenured Women Battle to Make It Less Lonely at the Top," *Science* 286 (November 12, 1999): 1272–1278.

8. Dean Robert Birgeneau's plane was canceled, and he never attended the meeting.

9. Robin Wilson, "An MIT Professor's Suspicion of Bias Leads to a New Movement for Academic Women," *Chronicle of Higher Education*, December 3, 1999, A16–A18.

10. Letter to Nancy Hopkins, February 24, 2008.

11. The group, now known as the MIT9, met again in 2004 and 2007; faculty members from these institutions met in the intervening years. The meeting in 2007 focused on underrepresented minorities.

12. "Subtle Discrimination Spurs MIT to Change," *San Francisco Chronicle*, March 24, 1999.

13. "Gender Bias on the Campus," *New York Times*, March 28, 1999.

14. Reynolds Holding, "When It Doesn't Pay to Confess Your Sins," *San Francisco Chronicle*, April 25, 1999.

15. Judith S. Kleinfeld, *MIT Tarnishes Its Reputation with Gender Junk Science* (Arlington, VA: Independent Women's Forum, December 1999).

16. Independent Women's Forum, <http://www.iwf.org/> (accessed June 20, 2008).

17. "Gender Bender," *Wall Street Journal*, December 29, 1999.

18. Ibid.

19. For an overview of the reports, see <http://web.mit.edu/faculty/reports/overview.html> (accessed December 24, 2008).

20. For the report of the Sloan School of Management, see <http://web.mit.edu/faculty/reports/som.html> (accessed December 24, 2008). In his introduction, the Dean of the Sloan School made the following comment: "By far the most surprising aspect of the Committee's work is its profoundly disturbing analysis of faculty experience. This analysis makes it inescapably clear that in our culture, men and women faculty with outwardly very similar careers are, in effect, working at two different Schools and that the women are at a much less congenial and supportive Sloan than the men."

21. The other MIT schools are Architecture and Planning; Engineering; Humanities, Arts, and Social Sciences; and the Sloan School of Management.

22. See <http://web.mit.edu/faculty/reports/overview.html>.

23. Letter from Provost Robert A. Brown to Nancy Hopkins, September 13, 2000.

24. When Clay became chancellor in June 2001, Professor Wesley Harris took his place as cochair.

25. Hopkins seems to have been the only person without an official position on the organizational chart to be made a member of the Academic Council. Her successors are now officially Associate Provosts of Faculty Equity.

26. This subcommittee was chaired by Professor Lorna Gibson. For the guidelines to ensure diversity in hiring, see <http://web.mit.edu/faculty/reports/FacultySearch.pdf> (accessed on December 24, 2008).

27. These policies had actually been proposed at an earlier time, but before the administration's commitment to the Diversity Council, they were not adopted.

28. This subcommittee was chaired by Professor Bailyn. The policies for family care for faculty are available at <http://web.mit.edu/faculty/benefits/familycare.pdf> (accessed December 24, 2008).

29. For a report on this conference, see the MIT News Office release from January 30, 2001, available at <http://web.mit.edu/newsoffice/2001/gender.html> (accessed December 24, 2008).

30. See Committee on Science, Engineering, and Public Policy, *Beyond Bias and Barriers: Fulfilling the Potential of Women in Academic Science and Engineering* (Washington, DC: National Academies Press, 2007). A subsequent equivalent study on

underrepresented minority faculty in science and engineering is currently in preparation.

31. Nancy Hopkins, "Diversification of a University Faculty: Observations on Hiring Women Faculty in the Schools of Science and Engineering at MIT," *MIT Faculty Newsletter* (March–April 2006).

32. Lotte Bailyn, "Academic Careers and Gender Equity: Lessons Learned from MIT," *Gender, Work, and Organizations* 10 (2003): 137–153, on 141.

33. Elements of the story of what led to the MIT Report are available in personal form in Nancy Hopkins's afterword to this chapter. See also Bailyn, "Academic Careers," especially 145–150; Nancy Hopkins, "MIT and Gender Bias: Following Up on Victory," *Chronicle of Higher Education* (June 11, 1999).

34. The following women faculty members agreed to sign the letter: from the Department of Biology, professors Sallie (Penny) Chisholm, Nancy Hopkins, Ruth Lehmann (now head of the Skirball Institute at New York University), Terry Orr-Weaver, Mary-Lou Pardue, and Lisa Steiner; from the Department of Brain and Cognitive Sciences, professors Susan Carey, Suzanne Corkin, Ann Graybiel, and Mary Potter; from the Department of Chemistry, professors Sylvia Ceyer and JoAnne Stubbe; from the Department of Earth, Atmospheric, and Planetary Sciences, professors Marcia McNutt (now head of the Monterey Bay Aquarium Research Institute), Paola Rizzoli, and Leigh Royden; and from the Department of Physics, Professor Millie Dresselhaus. Eleven of these women are members of the National Academies.

35. Email from Robert Birgeneau to the author, April 27, 2008.

36. The committee interviewed all the senior women, but in order to protect the junior women, they were asked if they would be willing to be interviewed; they could choose who would conduct the interview, and whether it would be alone or in a group.

37. By 1998, three new people—two women and one man—had been added to the committee: Professor Sylvia Ceyer from the Department of Chemistry; Professor Jacqueline Hewitt, recently tenured in the Department of Physics; and Professor Kip Hodges from the Department of Earth, Atmospheric, and Planetary Sciences.

38. The new chair of the faculty was Bailyn.

39. The *Faculty Newsletter* was established in 1988 in response to an unwelcome decision by the administration. In the "zeroth" edition, Professor Vera Kistia-

kowsky noted that the newsletter would be a "channel for the exchange of information between faculty members and for the discussion of problems at MIT since neither *Tech Talk* nor the faculty meetings serve these purposes." The *Faculty Newsletter* has done this ever since.

40. Bailyn, "Academic Careers."

41. Ibid., 149.

42. Christina Hoff Sommers, "Why Can't a Woman Be More Like a Man?" *American*, March–April 2008, available at <http://www.american.com/archive/2008/march-april-magazine-contents/why-can2019t-a-woman-be-more-like-a-man> (accessed December 24, 2008).

43. Professor Paula Hammond of the Department of Chemical Engineering is leading this effort. The report was released in January 2010, and is available at <http://web.mit.edu/provost/raceinitiative>.

44. The Associate Provosts for Faculty Equity are professors Barbara Liskov and Wesley Harris, both from the School of Engineering.

NANCY HOPKINS

Because of my involvement in what came to be known as the MIT Report and its aftermath, Professor Bailyn asked me to add a few personal comments to her account of how this report came to be written and the impact it had. I want to comment on the women whose professional experiences formed the basis of this report, the importance of data gathering for solving social problems, the role that Bailyn herself played in the outcome of this story, the necessity of having leaders—Dean Robert Birgeneau and President Charles Vest—who were willing to lead, the extraordinary progress for women in science that resulted from the report, and finally, prospects for a time when gender in science will no longer be a pervasive issue.

When I joined the MIT faculty thirty-five years ago, I, like most young women today, thought that gender discrimination was a thing of the past. I believed that the great civil rights and women's rights movements as well as the affirmative action laws of the late 1960s and early 1970s had eliminated gender discrimination by requiring universities to hire women to their faculties. I assumed that the only reason there were still so few women on university faculties, particularly science faculties, was that for most women, in an era before affordable day care, amniocentesis, or in vitro fertilization, raising children and being a top-notch scientist were incompatible activities. I also firmly believed that science was entirely merit based. Thus, if a woman was willing to forgo having children, willing to work hard, and if she made significant discoveries, she would not experience any more obstacles to a successful scientific career than a man.

I was among the first ten women hired to the MIT School of Science faculty of about 270 total members. It took me fifteen years of watching how other women faculty members were treated to realize that I had been wrong: I came to see that when women and men made discoveries of equal importance, the women were neither valued nor rewarded equally. Remarkably, it took me another five years to realize that this was probably even true for me. Previously, I had managed to convince myself that I was the one exception. The reality proved to be devastating. Furthermore, I assumed for a long time that I was the only woman scientist who had come to understand gender bias. After all, few women had tried to be scientists at elite universities, none who succeeded spoke about gender bias, and if it was so difficult for me to understand it, what were the chances that others had?

In 1994, however, when I was a full professor with tenure, I reached a point where I could no longer tolerate the frustrations of being unable to get the space and resources that were essential for my research, and that were readily available even to assistant professors just starting out. It was at that point that I asked my colleague Professor Mary-Lou Pardue to read a letter I had drafted to send to MIT President Vest asking for his help. To me, Pardue was a revered role model because of the extraordinary scientific contributions she had made in her career. Sharing my concerns with her was the hardest as well as most important moment in this whole saga. I feared she would think badly of me for blaming my difficulties on gender bias. Instead, Pardue read my letter, laid it down on the small table in Rebecca's Café, where we were having lunch, and said, "I'd like to sign this letter, and I think we should go and see President Vest." That moment changed my life and made possible all that followed. Years of self-doubt, frustration, and anger evaporated in that instant. I also think we both realized in that moment that together we really had a chance to address the problem. One woman complaining can easily be dismissed, but two women could not be easily brushed aside.

Rather than send my letter to Vest, I worked over that summer with Pardue and with our colleague Professor Lisa Steiner to poll the remaining fourteen tenured women faculty members in the six departments of the School of Science to determine if they had seen or experienced the subtle but damaging exclusion and bias that we had, and if so, whether they would be willing to join us in asking the administration to document the problem so that it could be better understood and then addressed. This proved to be the beginning of the most rewarding personal experience that I had at MIT. The fifteen women who joined me that summer in signing a letter to the Dean of Science, Birgeneau, were the most remarkable group of people that I've met in my career. They were pioneers, many of whom had achieved in science at the highest levels despite obstacles that often made their jobs and lives extraordinarily difficult. Along with brilliance, they shared an overwhelming passion for research as well as unusual tenacity and self-reliance. Most important for our mutual cause, despite their realistic concerns that they might be seen as complainers and thus damage their professional reputations, the women chose to work together (although in confidence) to improve the working environment for themselves, and they hoped, for future generations of women scientists. Today, fourteen years after we first met, I remain close friends with many of these extraordinary women.

Initially, Birgeneau did not want me to chair the committee that he established in 1995 and charged with studying the status of women faculty members in the School of Science: he feared that I might be too "radical." The other women, however, realized that I was willing to do the time-consuming work that lay ahead, so they declined to serve as chair, leaving the dean no option except to choose me. I soon put Birgeneau's fears to rest by showing him that I would always only represent the consensus of all sixteen women. When our committee first met, some women were interested primarily in obtaining numerical data. But others, especially ecologist Penny Chisholm and biochemist JoAnne Stubbe, recognized that individual women's stories were as essential and valid a form of data in explaining the nature of gender discrimination as the "numbers." Indeed, it was the two types of data together that helped the dean and President to understand the role that gender bias had played in these women's careers and lives.

As the chair of the first "committee on the status of women faculty in science," at first I found it difficult to obtain the numerical data needed to assess the status of female versus male faculty members, and assess the accuracy of the stories and perceptions that women told us. Fortunately, the dean had insisted that we include three powerful men on our committee. One day, one of them, Bob Silbey, then department head of chemistry, took me to Birgeneau's office, pounded on the dean's coffee table with his fist, and demanded that the dean give me the data. This proved to be a critical turning point. Reliable data, along with the professional assessments of quality that underlie all academic evaluations, are essential for tracking equity. Even today as I travel and speak on these issues, I find women grappling with administrations that refuse them access to the data necessary for this type of work. To my mind, this is a mistake that generates mistrust of the administration among the faculty and prevents institutions from rapidly progressing on gender equity problems.

Our committee's collaboration with Birgeneau was enormously productive. Although only some two dozen faculty participated in this process, it was the first time that women had dared to speak openly about sensitive issues—for example, the stigma attached to asking for a leave or tenure extension to have a child, the embarrassment of being omitted from group grants that included all other faculty members in one's field, or the discouragement on finding that one has been seriously underpaid for years. Jerry Friedman, a former department head of physics and Nobel Laureate who served on our committee, pressed me to put our findings into writing as soon as possible so that the dean could correct the inequities we found. I did so, and Birgeneau rapidly responded. The end result was to restore a sense of fairness among the women faculty members and make their research easier. It is impossible to overstate how significant the dean's actions were for many of the women involved. Yet we knew that if this powerful advocate were to leave MIT, the progress that had been made would end. I myself had spent about twenty hours per week for two years carrying out the committee's work and writing the 150-page confidential committee report. Importantly, in 1997 the dean appointed a second committee to continue the investigation, and Professor Molly Potter agreed to chair it. The findings of both committees, though, remained virtually unknown to the wider MIT faculty until 1999.

In 1999, Lotte Bailyn, who knew of our work, was chair of the MIT faculty. Bailyn also understood that the problems of gender bias that we had identified were not limited to either the fields of science or elite universities. Rather, they were reflections of women's status in society and hence pandemic in all of higher education. It was she who pushed the women faculty committee and the administration to summarize the processes that we had used along with the findings we had made, and then release them to the public. Without this, I suspect there would have been little lasting progress for women in science at MIT.

After I had drafted the public version of the report, Potter and Bailyn edited it, adding the often-quoted sentence that defines twenty-first-century discrimination as "a pattern of powerful but unrecognized assumptions and attitudes that work systematically against women faculty even in the light of obvious good will." I ran the draft by the women faculty members while Bailyn took it to Birgeneau and Vest. I can never forget the moment when Vest's now-famous comment appeared on my computer screen: "I have always believed that gender discrimination is part perception, part reality. True, but I now understand that reality is the greater part of the balance." I had always believed I would go to my grave before almost anyone, and certainly any college president, would come to understand the essentially invisible problem that had damaged the lives of so many generations of professional women. But now the President of MIT had come to realize it, and was decent and brave enough to acknowledge it, despite the obvious political and legal hazards of doing so. Birgeneau added powerful comments of support.

As Bailyn described above, the MIT Report ended up on the front pages of the *Boston Globe* and *New York Times*. A few weeks later, I found myself at the White House. Although Birgeneau was supposed to join me there, his plane was canceled, so I went alone. President Clinton and the first lady shook my hand, and said, "Thank you on behalf of the nation." They both gave speeches about how other institutions should follow MIT's lead and look at the data from deep inside their own institutions. As I flew home a few hours later, on a perfect spring day, I had to pinch myself. Had this really happened? The tenured women faculty in the School of Science had gone from meeting and working in near secrecy five years earlier to national visibility, including public endorsement by the President of the United States.

After the MIT Report hit the front page of the *New York Times*, I was overwhelmed by email, primarily from grateful women, and by calls from the press. For a year and a half, I spoke with the press at least five days a week, often to multiple reporters every day, and even now, almost ten years later, I am usually called at least every week with requests for information or my comments on related issues. I have been enormously impressed by the quality of the press coverage and believe that the media get credit for rapidly educating the public about this issue. A possible side benefit early on was remarked on by a reporter who told me, "Without us, you'd be dead by now." He referred both to the fact that some colleagues probably did not understand or appreciate the MIT Report, and the right-wing anger and backlash that he assumed would occur. He was probably correct that the enormous visibility of the MIT Report protected me from these reactions to some extent. Most interesting to me, however, were the women I met when lecturing on this issue at more than a hundred academic institutions. These women would begin their stories, and I could finish them before they did because the experiences of bias in all academic fields and institutions are so similar.

How far have we come in terms of gender equity for women scientists, and how far is there to go? Recently I reread the 1995–1996 report we made to Birgeneau. It felt to me as if it could have been written a hundred years ago, because there has been so much change and progress since then. Essentially, all the recommendations that our committee made—most of which seemed radical then—have not only come to pass but also have been institutionalized at MIT and, indeed, many universities in the United States. Perhaps most important, it became possible to discuss these issues openly, ending the painful years of isolation and frustration that many women experienced. Also enormously significant, women faculty were appointed to high positions in the administration, including today the presidency of MIT. As a result, most women faculty members now have knowledge of how the system works.

Despite these accomplishments, however, there are two areas where we have not yet succeeded, in my opinion. First, the numbers of women faculty members remain low. Why this is and even how much it matters are not clear to me, but the issue continues to merit serious analysis and attention. Second, and more critical, the professionally damaging and disheartening marginalization of women that we documented in 1995, while less extreme today, still occurs. This problem is hard to fix because, as Bailyn first taught me, it really reflects the status of women in our society. We should perhaps be more concerned that there has never been a woman president of the United States than that, as of 2008, there has never been a tenured female professor of math at Harvard (and only four at MIT). Only when it is no longer front-page news that a woman has done this or that, only when the number of women in Congress reflects the number of women in the country, only when the number of women on the MIT faculty more closely reflects the number of women we train, and only when women and men interact professionally without regard to gender, will we have achieved equality. I hope MIT will continue to be a leader toward these goals.

SUSAN HOCKFIELD

On the cusp of MIT's sesquicentennial, the Institute's distinctive character and indelible achievements might tempt one to view its history as inevitable, a straightforward tale of unidirectional growth. Yet the chapters here reveal a richer, more complicated narrative. Choices that in retrospect seem to have been simple and obvious appeared in their time as crises, dilemmas, and battles to find the right path. By responding to such challenges with creativity and bold experiment, the community of MIT gradually built the Institute we know today.

Inheriting this dynamic legacy, we approach MIT's 150th anniversary facing important decisions of our own. Fortunately, from the formative choices described in these pages, important lessons emerge. At least two stand out clearly. The first is that MIT's founding ideals have served it well. For example, in launching the Institute, our founder William Barton Rogers favored fundamental scientific principles and direct experimentation over "the minute details and manipulations of the arts"; decades later, adherence to this ideal saved MIT from reducing itself to an industrial training school. Similarly, Rogers believed deeply in a mission of service to society. This fundamental value inspired MIT's critical contributions during and after World War II.

A second lesson is that in its institutional thinking as in its most advanced research, the Institute should reach for new frontiers. The historic institutional choices outlined in this book were made by MIT, for MIT; yet because of the Institute's stature and impact, several moments that were decisive for MIT bent the larger arc of history as well. MIT's model, which in the 1940s put into form the concept of the federally funded research university, triggered historic advances in America's security, health, innovation, and prosperity. Since the 1960s, MIT's bold embrace of industrial partners as conduits to key research problems encouraged the work of faculty and student entrepreneurs, and has helped universities serve as engines of economic growth. And in the 1990s, by openly recognizing and working to redress ways that it had inadvertently disadvantaged women scientists, MIT opened a crucial national conversation.

Today, MIT faces a range of vital questions, from how we integrate our physical campus with virtual resources as we educate our students to how we shape our institutional engagements around the world. We return, as in the past, to the touchstone of first principles and find inspiration in our founder's insistence on new frontiers. For instance, in 2006, MIT made a historic commitment of talent and resources to tackle the vast global problem of sustainable energy. The MIT Energy Initiative currently reflects our enduring belief in the importance of basic research and unfettered experimentation, our focus on hands-on learning, and our commitment to serving society—values instilled by Rogers himself. This initiative not only pursues new frontiers in basic and applied research but its definitive energy policy reports also frame the national debate, and it defines ambitious educational standards for a new generation of energy pioneers.

MIT had no official archives until 1961, when the Institute's 100th anniversary prompted a recognition of our important and enduring contributions. If the decisions described in these pages had turned out differently, we might not be preparing to celebrate our 150th anniversary today. Our preparations for MIT's sesquicentennial remind us of our enormous responsibility to make decisions in the present that will serve the future as well as those of the past serve us today.

Gray House
Cambridge, Massachusetts

LOTTE BAILYN is Professor of Management and T. Wilson (1953) Professor of Management emerita at MIT's Sloan School of Management. For the period 1997–1999 she was chair of the MIT faculty, and during 1995–1997 she was the Matina S. Horner Distinguished Visiting Professor at Radcliffe's Public Policy Institute. Bailyn is an authority on workplace conditions, particularly as they affect the careers and lives of technical and managerial professionals. Her research deals with the relation of organizational practice to employees' personal lives, with an emphasis on gender equity in business organizations and academia. She was the principal investigator for the research effort of the recently completed MIT Initiative on Faculty Race and Diversity. Among her publications are *Breaking the Mold: Women, Men, and Time in the New Corporate World* (Free Press, 1993) along with its new and fully revised edition *Breaking the Mold: Redesigning Work for Productive and Satisfying Lives* (Cornell, 2006), and *Beyond Work-Family Balance: Advancing Gender Equity and Workplace Performance* (Jossey-Bass, 2002), of which she is a coauthor.

DEBORAH DOUGLAS is the Curator of Science and Technology at the MIT Museum. She has also held positions at the National Air and Space Museum, the NASA Langley Research Center, and the Chemical Heritage Foundation, and taught as an adjunct assistant professor at Old Dominion University. Douglas has curated twenty exhibitions on a wide variety of MIT-related science and technology topics, including "Mind and Hand: The Making of MIT Scientists and Engineers," and is also the curator for the Institute's 150th-anniversary exhibition opening in 2011. A specialist in aerospace history, she is the author of *American Women and Flight since*

1940 (University Press of Kentucky, 2004) as well as several essays, articles, and reviews.

JOHN DURANT trained in natural sciences and the history of science at the University of Cambridge. After more than a decade in university continuing education (first at the University of Swansea in Wales, and then at the University of Oxford in England), in 1989 he was appointed Assistant Director and Head of Science Communication at the Science Museum in London, and Professor of Public Understanding of Science at the Imperial College in London. In 2000, he took up the position of chief executive of At-Bristol, a new independent science center in the west of England. He came to MIT in 2005, where he holds a joint appointment as Director of the MIT Museum and an adjunct professor in the Program in Science, Technology, and Society. Durant is active in research, teaching, and practice in the interdisciplinary field of public engagement with science and technology.

SUSAN HOCKFIELD has served as the sixteenth president of MIT since December 2004. A noted neuroscientist whose research focused on the development of the brain, she also holds a faculty appointment in the Department of Brain and Cognitive Sciences. During her tenure as president, MIT has capitalized on its strength in engineering as well as in the physical and life sciences to produce groundbreaking interdisciplinary research on topics from cancer to autism to AIDS. She has also championed the faculty-led MIT Energy Initiative, which supports pioneering, cross-disciplinary research, policy, and education to help address the global challenge of sustainable energy; the initiative's definitive reports on key areas of energy policy have helped frame the national debate. Hockfield has also helped MIT cultivate strong international engagements in research and education, from China and India to Europe and the Middle East. She came to MIT from Yale University, where she joined the faculty in 1985 and was later named William Edward Gilbert Professor of Neurobiology. At Yale, Hockfield emerged as a strong, innovative leader, first as dean of its Graduate School of Arts and Sciences, and then as provost. After earning a BA in biology from the University of Rochester and a PhD from Georgetown University at the School of Medicine, Hockfield was an NIH postdoctoral fellow at the University of California at San Francisco and later joined the scientific staff at the Cold Spring Harbor Laboratory.

NANCY HOPKINS is the Amgen, Inc. Professor of Biology at MIT. She is a member of the National Academy of Sciences, a fellow of the American Academy of Arts and Sciences, and a member of the Institute of Medicine of the National Academies. Hopkins obtained a BA from Radcliffe College in 1964, and then a PhD in molecular biology and biochemistry from Harvard University in 1971. As an undergraduate, she was inspired to become a scientist after hearing a lecture by James D. Watson, the codiscoverer of the structure of DNA. In 1974, Hopkins joined the MIT faculty as an assistant professor in what is now the Koch Institute for Integrative Cancer Research at MIT. Hopkins's lab is known for identifying a significant fraction of the genes essential for early vertebrate development using the zebrafish as a model system. Today her lab studies mechanisms by which genetic alterations predispose zebrafish to cancer. Hopkins codeveloped and taught the first first-year undergraduate biology course required of all MIT undergraduates. She was appointed chair of the first Committee on Women Faculty in the School of Science in 1995. In 2000, Hopkins was appointed cochair with Provost Robert Brown of the first Council on Faculty Diversity and served on the Academic Council.

DAVID KAISER is Germeshausen Professor of the History of Science and Director of MIT's Program in Science, Technology, and Society, and Senior Lecturer in MIT's Department of Physics. He completed PhDs in physics and the history of science at Harvard University. Kaiser is author of the award-winning book *Drawing Theories Apart: The Dispersion of Feynman Diagrams in Postwar Physics* (University of Chicago Press, 2005), which traces how Richard Feynman's idiosyncratic approach to quantum physics entered the mainstream. His edited books include *Pedagogy and the Practice of Science: Historical and Contemporary Perspectives* (MIT Press, 2005). *How the Hippies Saved Physics: Science, Counterculture, and the Quantum Revival* (W. W. Norton) was published in 2011. He is presently completing *American Physics and the Cold War Bubble* (University of Chicago Press). His work has been featured in such venues as *Science*, *Harper's*, and *Scientific American*, on NOVA television programs, and on National Public Radio's *Science Friday*. Honors include awards from the American Physical Society, the History of Science Society, the British Society for the History of Science, and MIT. He has also received several teaching awards from Harvard and MIT.

CHRISTOPHE LÉCUYER, a graduate of the Ecole Normale Supérieure, earned his PhD in history from Stanford University. He taught at Stanford and the University of Virginia, and is currently a principal economic analyst at the University of California. He published extensively on the history of electronics, instrumentation, and high-tech manufacturing. He is the author of *Making Silicon Valley: Innovation and the Growth of High Tech, 1930–1970* (MIT Press, 2006) and (with David C. Brock) *Makers of the Microchip: A Documentary History of Fairchild Semiconductor* (MIT Press, 2010).

STUART W. LESLIE has taught the history of science and technology at the Johns Hopkins University since 1981. He has published *Boss Kettering* (Columbia University Press, 1983), a biography of General Motors executive and engineer Charles Kettering, and *The Cold War and American Science: The Military-Industrial-Academic Complex at MIT and Stanford* (Columbia University Press, 1993), a study of U.S. science and engineering education in the postwar period. Leslie has also written extensively on the rise and fall of science regions in the United States and abroad, and on efforts to emulate U.S. models of science and engineering education in the developing world. His current research focuses primarily on the history of laboratory design and architecture, including the work of I. M. Pei, Louis Kahn, and Eero Saarinen, part of a larger project titled *The Architects of Modern Science*.

BRUCE SINCLAIR did his undergraduate work at the University of California at Berkeley, where he majored in history. He became interested in industrial history as a Hagley Fellow at the University of Delaware, and served as the founding director of the Merrimack Valley Textile Museum, subsequently renamed the Museum of American Textile History. He received his PhD in the new graduate program in the history of technology that Melvin Kranzberg established at Case Institute, and went on to teach first at Kansas State University and then for the next twenty years at the University of Toronto, where he was director of the Institute for the History and Philosophy of Science and Technology. When the Georgia

Institute of Technology created the Melvin Kranzberg Professorship in the History of Technology, Sinclair was named the first occupant of that chair. Most of his research and writing has centered on technical education and the institutional history of engineering. His first book, *Philadelphia's Philosopher Mechanics: A History of the Franklin Institute, 1824–1865* (Johns Hopkins University Press, 1974), was awarded the Dexter Prize by the Society for the History of Technology. He has also written a history of the American Society of Mechanical Engineers, edited a history of Canadian technology, and edited a collection of essays about African Americans and their experience with technology. His chapter in this volume grew out of a fascination with the tangled relations between MIT and Harvard. Sinclair served as president of the Society for the History of Technology, which subsequently awarded him its Leonardo da Vinci Medal. His principal deviation from an otherwise-reasonably sane life has been a weakness for old wooden boats.

MERRITT ROE SMITH is the Cutten Professor of the History of Technology in the Program in Science, Technology, and Society and the History Faculty at MIT. His research focuses primarily on American manufacturing and industry from the 1790s to World War I. He is the author or editor of six books, the most recent being *Inventing America: A History of the United States* (2nd ed., W. W. Norton, 2006). Among his many recognitions are the presidency of the Society for the History of Technology, book awards from the Organization of American Historians and the History of Science Society, a Regent's Fellowship from the Smithsonian Institution, the Leonardo da Vinci Medal from the Society for the History of Technology, and elected fellowships in the American Academy of Arts and Sciences, the American Association for the Advancement of Science, and the Massachusetts Historical Society. He is currently working on a book about technology and its implications during the American Civil War era, and is coeditor of a forthcoming volume titled *Reconceptualizing the Industrial Revolution* (MIT Press). In addition to editing the Johns Hopkins University Press series in the history of technology, Smith serves on the national advisory boards of the Thomas A. Edison Papers Project (Rutgers University), the American Precision Museum, and the public television series *The American Experience*.

Note: The letter *f* following a page number denotes a figure.

Printed in the United States
by Baker & Taylor Publisher Services